Introduction to Thermodynamics

To Mandy

Introduction to Thermodynamics

Keith Sherwin

Senior Lecturer in Thermodynamics and Fluid Mechanics,
University of Huddersfield

CHAPMAN & HALL

London · Glasgow · New York · Tokyo · Melbourne · Madras

Published by Chapman & Hall, 2–6 Boundary Row, London SE1 8HN

Chapman & Hall, 2–6 Boundary Row, London SE1 8HN, UK

Blackie Academic & Professional, Wester Cleddens Road, Bishopbriggs, Glasgow G64 2NZ, UK

Chapman & Hall Inc., 29 West 35th Street, New York NY10001, USA

Chapman & Hall Japan, Thomson Publishing Japan, Hirakawacho Nemoto Building, 6F, 1-7-11 Hirakawa-cho, Chiyoda-ku, Tokyo 102, Japan

Chapman & Hall Australia, Thomas Nelson Australia, 102 Dodds Street, South Melbourne, Victoria 3205, Australia

Chapman & Hall India, R. Seshadri, 32 Second Main Road, CIT East, Madras 600 035, India

First edition 1993

© 1993   Keith Sherwin

Typeset in 10.5/12pt Times by Expo Holdings, Malaysia
Printed in Great Britain by Clays Ltd, Bungay, Suffolk

ISBN 0 412 47640 1

A catalogue record for this book is available from the British Library

Library of Congress Cataloging-in-Publication data

Sherwin, Keith.
    Introduction to thermodynamics/Keith Sherwin. — 1st ed.
       p.   cm
    Includes index.
    ISBN 0-412-47640-1
    1. Thermodynamics.   I. Title.
TJ265.S494   1993
621.402'1—dc20   92-38960
                                                                       CIP

Contents

Preface

As the title implies, this book provides an introduction to thermodynamics for students on degree and HND courses in engineering. These courses are placing increased emphasis on business, design, management, and manufacture. As a consequence, the direct class-time for thermodynamics is being reduced and students are encouraged to self learn.

This book has been written with this in mind. The text is brief and to the point, with a minimum of mathematical content. Each chapter defines a list of aims and concludes with a short summary. The summary provides an overview of the key words, phrases and equations introduced within the chapter.

It is recognized that students see thermodynamics as a problem-solving activity and this is reflected by the emphasis on the modelling of situations. As a guide to problem solving, worked examples are included throughout the book. In addition, students are encouraged to work through the problems at the end of each chapter, for which outline solutions are provided.

There is a certain timelessness about thermodynamics because the fundamentals do not change. However, there is currently some debate over which sign convention should apply to work entering, or leaving, a thermodynamic system. I have retained the traditional convention of work out of a system being positive. This fits in with the concept of a heat engine as a device that takes in heat and, as a result, produces positive work.

I wish to thank the following organizations for their permission to reproduce the illustrations listed below: the Keighley and Worth Valley Railway (Figure 1.4); the Trustees of the Science Museum (Figure 1.5); Rolls Royce plc (Figures 1.7 and 1.8); the Ford Motor Company Ltd (Figure 8.8); and National Power plc (Figure 10.12). In addition, I would like to thank ICI Chemical Products for kindly providing information on their alternative Refrigerant-134a.

Many people have helped with this book: Dr H. Barrow, Dr G. McCreath and Prof H. Rao with wide-ranging discussions on thermodynamics, life and the universe; Dr Y. Mayhew for his permission to use the data on steam and Refrigerant-12 given in Appendix A. The complete draft was read by Dr G. I. Alexander and I am grateful for his constructive comments. Finally, I wish to dedicate this book to my wife as a small tribute to her help and encouragement during its preparation.

Keith Sherwin

List of symbols

The following symbols have been used throughout the book:

A	Area
C_P, C_v	Specific heat
COP	Coefficient of performance
F	Force
g	Gravitational constant
h	Specific enthalpy
$\bar{h}_c$	Enthalpy of combustion
$\bar{h}_f$	Enthalpy of formation
h_c	Convective heat transfer coefficient
H	Total enthalpy
k	Thermal conductivity
KE	Kinetic energy
m	Mass
$m\dot{}$	Mass flow rate
n	Number of kmol
M	Molecular weight
P	Pressure
PE	Potential energy
q	Heat
$\bar{q}$	Heat transferred during combustion
Q	Rate of heat transfer
r	Compression ratio
r	Radius
R	Gas constant
R_o	Universal gas constant
s	Specific entropy
t	Time
T	Temperature
u	Specific internal energy
U	Overall heat transfer coefficient
v	Specific volume
V	Velocity
$\mathscr{V}$	Volume
w	Work
W	Power
x	Dryness fraction
x, y, z	Distance
γ	Ratio of specific heats C_P/C_v
ϵ	Emissivity
η	Efficiency
σ	Stefan–Boltzmann constant
ϕ	Relative humidity
ω	Specific humidity

Introduction $\boxed{1}$

1.1 AIMS

- To define thermodynamics as an engineering science.
- To provide an overview of the development of thermodynamics.
- To provide a brief historical review of the development of steam engines.
- To provide a brief historical review of the development of internal combustion engines.
- To introduce the SI system of units.
- To introduce the basic dimensions used within the SI system.
- To introduce the derived units for force, energy and power.
- To define pressure and explain the difference between gauge and absolute pressure.
- To define the Celsius and absolute scales of temperature.
- To introduce the kmol as the molecular mass of gases.

1.2 THERMODYNAMICS

The problem with definitions is that they are either so broad as to be virtually meaningless, or so hedged round by qualifying statements that they leave the reader totally confused. Defining thermodynamics is no exception.

Thermodynamics can be defined as a science dealing with energy. However, to someone having no prior knowledge of thermodynamics this means nothing. That it is a science is true, but a science simply represents a sum of knowledge. Section 1.3 gives an overview of the ideas and theories that go to make up the scientific content of present day thermodynamics. Thermodynamics also involves the use of that knowledge to create practical devices in the form of engines.

The application of thermodynamics knowledge takes place within engineering. It is engineers who design and manufacture steam plant for power stations or turbo-fan engines to propel modern airliners. In order to design these devices they must have a sound knowledge of, and be able to apply, thermodynamics.

Thermodynamics, within the context of this book, can be considered to be an engineering science. It forms an essential part of the education of an engineer since the conversion and use of energy are essential for any modern society. All the domestic accoutrements of modern living, such as electric light at the flick of a switch, storage of food in a refrigerator, personal transport in the form of a motor car, depend on the thermodynamic applications of energy.

Energy comes in many forms and it is possible to have mechanical energy that is totally independent of thermodynamics. A rock rolling down a hillside gains kinetic energy as a result of losing potential energy. This is a change of energy, but one that does not need a thermodynamic process to take place.

Thermodynamics is largely concerned with the conversion of heat into work. What is generally referred to as 'heat' is, in fact, thermal energy. Thermal energy is released by the burning of fossil fuels or within nuclear reactors. Some part of this thermal energy may be transferred as heat in order to be converted into work. The work may be in the form of power from a rotating shaft or in the form of a propulsive force produced by an aircraft engine. It is, therefore, possible to give a more relevant definition of thermodynamics – as an engineering science dealing with conversion of heat energy into mechanical work.

This definition is certainly an improvement on that given earlier and is adequate to give an overall view of thermodynamics. However, it does not cover all aspects of the subject. In order to analyse the conversion of heat into work it is necessary to understand the behaviour of the fluids used in thermodynamics situations. Also, heat is a form of energy that is transferred only due to a temperature difference. It is essential to have an understanding of heat transfer. Therefore, in the subsequent chapters of this book, there is not only a discussion of heat engines and the basic laws that apply to them, but also chapters devoted to the properties of fluids and to heat transfer.

1.3   DEVELOPMENT OF THERMODYNAMICS

By the beginning of the 20th century much of the science of thermodynamics presented within this book had already been developed. Before discussing the ideas and theories that had evolved during the preceding three centuries that led to this body of knowledge, it is instructive to briefly look at the way that ideas are developed.

Within engineering design, innovation is aided by considering the creative process by which ideas are initiated (Whitfield, 1975). What is true for the creative processes within design is also true for the creation of ideas within thermodynamics.

Archimedes is reputed to have thought of his principle, of a body in a liquid displacing an amount of liquid equal to the body's weight, whilst having a bath. The story goes on that he jumped out of the bath shouting 'eureka', meaning 'I have found it'. This is a delightful story but in practice the development of ideas rarely comes from a sudden flash of inspiration but from the slow build up of knowledge from preceding ideas.

Thermodynamics is a subject that has been developed from painstaking work, observing, experimenting and creating theories to fit the known facts. Although many of the ideas required deep intellectual insight, nevertheless, the laws and relationships that are now available did not result from sudden 'eureka' changes in thinking but from a gradual development of the subject.

The year 1600 serves as a useful starting point for a brief outline of some of the development of thermodynamics. At that time knowledge of fluids was very rudimentary, harking back to the Greek philosophers' concept that all things were composed of four elements: air, fire, earth and water. However, there was some understanding of temperature as a measure of 'hotness' or 'coldness'. Galileo, in 1592, is credited as being the first person to construct a simple thermometer. It was open to the atmosphere and, therefore, subject to atmospheric pressure. The scale was purely arbitrary but it represented the first practical device for registering changes of temperature. The 17th century saw the gradual development of the thermometer until a sealed unit, similar to those used today, was available by 1650.

One of the most significant developments of the 17th century was the realization that the atmosphere has pressure. As mines became deeper, in order to find the reserves of coal or minerals, pumping of water from the mine became increasingly more important. It was found that a suction pump could not raise water above a height of about 10 m. It was Torricelli, a pupil of Galileo, who in 1644 announced that the reason for this was that the pressure of the atmosphere was equivalent to a column of water 10 m high. From the realization that the atmosphere had pressure came the idea that by working against a vacuum, the atmosphere could exert great forces. This led to the atmospheric engine, of the type designed by Newcomen, described in the next section.

In England the investigation of scientific phenomena was given impetus by the creation of the Royal Society in 1660. Although the investigations spread over the whole range of what are now effectively chemistry and physics, some aspects were directly applicable to thermodynamics. For example, in 1662 Robert Boyle developed his famous law that for an 'elastic' fluid, i.e. a gas, at constant temperature the variation of pressure with volume follows the relationship

pressure × volume = constant

This relationship was quoted a century later by James Watt in his steam engine patent.

During the 18th century several significant developments took place. In 1701 Isaac Newton presented the law of cooling, which is the basis of all convective heat transfer analysis, given in Chapter 12. It was realized that some form of standardized temperature scale was necessary instead of the arbitrary scales used before. Several different scales were created but the two still in use today are the Fahrenheit scale of 1724 and the Celsius scale of 1742 (This was commonly referred to as the centigrade scale as it had 100 graduations from the freezing point to the boiling point of water. In 1948 it officially became the Celsius scale again.) In 1787 Jacques Charles developed a relationship for the behaviour of a gas under constant pressure conditions:

volume/temperature = constant

which, when combined with Boyle's law, gives the 'equation of state' for a perfect gas, presented in Chapter 4.

The 19th century saw the development of thermodynamics into a mature science. The chemical nature of matter as being composed of atoms was firmly established by the year 1808. In 1811, Avogadro put forward his hypothesis that equal volumes of different gases contain an equal number of molecules providing that they are at the same pressure and temperature.

At the beginning of the century the concept of heat was widely debated. One popular theory was that heat was a fluid, without colour or weight, called 'caloric'. When an object became full of caloric it became 'saturated', a term still used today. From this concept of heat, Carnot developed the idea that there must be a flow of waste heat from an engine. In 1824 he formulated what is now known as the second law of thermodynamics, that a heat engine must always work with an efficiency of less than 100%.

Careful experimental work carried out by Joule in the 1840s showed that heat was not a fluid, but a form of energy and that, as such, it could be converted to other forms of energy. This led to the first law of thermodynamics which is really a statement of the conservation of energy. More particularly, it states that heat can be converted to work, as in an engine, or that work can be converted to heat, as a result of friction. The first and second laws of thermodynamics are the basis of all engineering thermodynamic analysis.

In 1848 William Thomson, later to become Lord Kelvin, used Carnot's conclusion regarding the efficiency of a heat engine to develop the concept of an absolute temperature scale. The absolute temperature scale based upon the degree Celsius is given the name 'Kelvin' in recognition of this work. Thomson is also reputed to have been the first person to have used the title 'thermodynamics' to define this branch of science.

To a Scottish engineer, William Rankine, goes the credit of writing the first textbook on engineering thermodynamics. Entitled *Manual of the Steam Engine and other Prime Movers* it was first published in 1859 and remained in print until the early years of the 20th century.

Table 1.1 Important stages in the development of thermodynamics

Date	Science	Technology
1600	Galileo's thermometer	
1700	Pressure measurement	
	Boyle's law	
	Newton's law of cooling	Savory engine
	Fahrenheit temperature scale	Newcomen engine
	Celsius temperature scale	
		Watt engine
1800	Charles' law	
	Avogadro's hypothesis	
	Carnot's Second law	
		'Rocket' locomotive
	Joule's First law	
	Absolute temperature scale	
	Rankine's textbook	
		Otto engine
		Steam turbine
1900		Motor car
		Jet engine

The important stages in the development of thermodynamics as an engineering science are presented in Table 1.1. For comparison, the key technological advances in the development of the heat engine are also listed.

1.4   DEVELOPMENT OF THE STEAM ENGINE

The history of the development of the steam engine up to 1900 is well covered (Derry and Williams, 1970). It is a story of the search for increasing efficiency in the use of fossil fuels to provide mechanical power. In some cases the development required prior knowledge from science, as it existed at the time. In other cases the development of engine technology was an impetus to scientific investigation.

What must also be borne in mind is that the development of engines depended on the technology of materials and manufacture that existed at the time. The building of the early Newcomen engine, described below, could not have taken place without the production techniques developed for the manufacture of cannons. The increase in steam pressures used within engines during the 19th century could not have been achieved without associated improvement in boiler manufacture, made possible by developments in the production of wrought iron and steel during that period. In the 20th century the development of a practical jet engine by Whittle would have been impossible without the availability of suitable high-temperature alloys.

Stemming from the work on atmospheric pressure during the seventeenth century, Savory devised a steam-driven machine for pumping water from

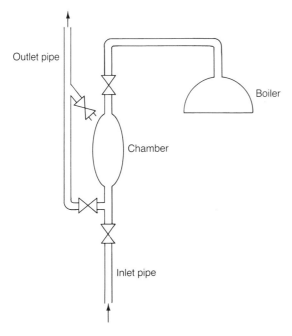

Figure 1.1 Diagram of the Savory engine.

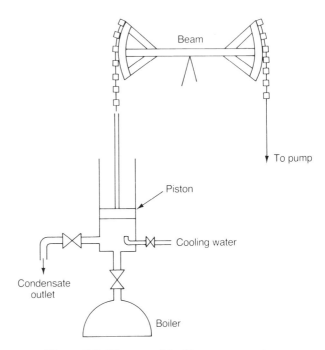

Figure 1.2 Diagram of the Newcomen engine.

mines. It was described as 'the Miner's Friend' when first patented in 1698 and had no working parts, except for hand operated valves. Figure 1.1 shows a schematic diagram of the device which worked by condensing steam in the chamber, using the partial vacuum created to suck water from the mine, then allowing steam to enter the chamber to force the water through the outlet pipe.

Independently, Newcomen developed an engine that used atmospheric pressure to drive a piston inside a cylinder. A schematic diagram of this 'atmospheric' engine is given in Figure 1.2. The operation of the engine was as follows.

1. Low pressure steam entered the cylinder, allowing the piston to rise.
2. After closing the steam valve, water was sprayed into the cylinder to condense the steam.
3. Atmospheric pressure forced the piston down against the partial vacuum formed in the cylinder, driving the pump through movement of the beam.
4. Condensate was drained from the cylinder and the cycle repeated.

Newcomen's first successful engine, with a cylinder diameter of 21 inches (0.53 m), was built in 1712 near Dudley Castle (then in Staffordshire, now West Midlands; a full-sized replica of this engine can be seen at the Black Country Museum at Dudley). The operation was made fully automatic and the engine could achieve a working rate of 12 strokes a minute. With such an intermittent action the Newcomen engine was well suited to working pumps but could not provide a continuous rotary motion. Eventually, examples of the Newcomen engine were used to pump water to reservoirs in order to drive water wheels for a rotary output.

The main problem with the Newcomen engine was the low operating efficiency, the ratio of the work output to the energy input of the fuel burnt. For the early engines this amounted to a fraction of 1%. Smeaton carried out an investigation of several Newcomen engines during the 1760s using a defined duty for comparison. The duty he defined was not an efficiency as such but a measure of the water pumped for the consumption of a bushel of coal. He found that those engines with the largest cylinder diameter performed best. This was due to the reduction of leakage around the piston compared to the steam in the cylinder. By building engines with an increased diameter, 52 inches (1.32 m), and using improved cylinder-boring techniques, he was able to achieve efficiencies of about 1%.

However, the main cause of inefficiency was the alternate heating and cooling of the cylinder during each stroke. When steam entered the cylinder, some would immediately be condensed in order to heat up the cylinder. This was realized by James Watt who, in 1769, patented the idea of a separate condenser. However, it was not until 1775 that the firm of Boulton and Watt started manufacturing improved Newcomen engines with condensers outside the cylinders. Watt's other improvements were the use of high-pressure steam instead of atmospheric pressure to drive the piston and then to enclose the end of the cylinder to make the engine double acting.

Watt's improvements brought about a considerable increase in the effective use of fuel and by the start of the 19th century, engines were operating with efficiencies of the order of 4–5%. A trend, during the early 1800s, towards higher steam pressures meant that engines could be made small enough for land transportation and that efficiencies improved still further. By 1850 steam pressure of 15 atmospheres had been achieved for stationary engines and efficiencies had risen to as high as 15%.

The first passenger-carrying steam railway was opened in 1830 between Liverpool and Manchester. For this railway, Robert Stephenson designed the *Rocket* locomotive shown in Figure 1.3. Although the boiler pressure was under 4 atmospheres, it set the pattern for future steam locomotive design by employing a multi-tubed boiler and directly coupling the pistons to the driving wheel. Gradual improvement of the steam locomotive during the 19th century resulted from increasing boiler pressures and, for some locomotive designs, a system of compounding in which the steam was partially expanded in a high-pressure cylinder before moving to low-pressure cylinders to complete the expansion. During the early years of the 20th century a further improvement

Figure 1.3 The *Rocket* steam locomotive.

Figure 1.4 British Railways class 4 locomotive, built in 1954.

resulted from the superheating of the steam entering the cylinders. Steam locomotives, as typified by that shown in Figure 1.4, remained in use on British railways until the 1960s but, even at best, the overall efficiency rarely exceeded 10%.

With any reciprocating steam engine there is a limitation to the power that can be achieved within a cylinder, irrespective of the steam inlet conditions. As the size of the cylinder and associated piston increase, the inertia of the moving parts restricts the speed at which it can rotate. No such limitation applies to steam turbines which were first developed in the 1880s. By 1884 Parson had built a steam turbine for driving a generator that ran at a speed of 18 000 rpm. The advantages of space saving and freedom from vibration over reciprocating steam engines, meant that steam turbines became increasingly used for applications requiring high-power outputs such as electrical generation and ship propulsion. In fact, the liner *Queen Elizabeth II* was powered by steam turbines when originally built.

However, the main disadvantage of any steam plant is the need for boilers and condensers. Observation of most power stations indicates that the landscape is dominated by rows of cooling towers, simply there to dissipate waste heat from the condensers. By using air as the working fluid instead of water this disadvantage can be partially overcome. In practice air is less dense than water or steam so, for a given mass flow rate, the increase in air volume means that engines working on air have to be larger than the equivalent steam turbine. For very high-power outputs, steam is still the most suitable working fluid. In modern power stations, steam driven turbo-generator sets are in use with outputs up to 1200 MW.

1.5 DEVELOPMENT OF INTERNAL COMBUSTION ENGINES

For low-power outputs, engines using air as the working fluid have an advantage and this was appreciated in the 19th century. However, the development of a practical engine working on air depended on the availability of a suitable fuel for combustion internally within the cylinder. This was solved with the widespread availability of coal gas. In 1859 a French engineer, Etienne Lenoir, built a gas engine that was designed along the lines of contemporary steam engines. Combustion was achieved by spark ignition using an electric induction coil. Performance was poor compared to equivalent steam engines, as the gas–air mixture was not compressed before combustion.

This defect was rectified in the Otto gas engine of 1876. This engine worked on the so-called 'Otto cycle', the constant volume air cycle described in Chapter 8. So successful were the Otto engines that the cycle was the basis of all subsequent four-stroke gas and petrol engines. Gas engines were clearly limited to stationary applications and it was the availability of liquid fuels that allowed the development of engines for propelling motor cars, boats and eventually aircraft.

The first practical petrol engine is attributed to Gottlieb Daimler. In 1885 he patented a single-cylinder engine, mounted vertically, in which ignition was achieved by means of a heated tube inserted into the engine cylinder head. Several of the earliest motor car engines used the same principle, but the hot tube was soon superseded by electric spark ignition.

Analysis of the Otto cycle shows that both power and efficiency improve with increasing compression ratio. The development of the petrol engine during the 20th century has centred around increasing the compression ratio, as a result of gradual improvement in the octane rating of fuels. Prior to 1940 many popular car engines were designed with side-valves, for ease of maintenance by owner–drivers. However, with a side-valve arrangement the compression ratio is limited by the need to provide space in the cylinder head to allow flow from the valves into the cylinder. Figure 1.5 shows a typical 1920s engine having side-valves and an 'up-draught' carburettor. For this engine, the compression ratio was under 5:1, compared to present-day engines with ratios of 10:1 achieved with an overhead-valve arrangement.

On the other hand, much higher compression ratios are required for compression-ignition engines. The Diesel engine was first manufactured in 1897 and was based upon ideal thermodynamic principles. Diesel reasoned that by using a high compression ratio, of the order of 20:1, the high temperature achieved by compressing the air in the cylinder would cause spontaneous combustion of the fuel. If the combustion was gradual it would take place at constant temperature, thereby eliminating the need for cooling. In practice this cannot be achieved, but the high compression ratios ensure an excellent operating efficiency.

As well as developments to reciprocating engines using air as the working fluid, the 20th century has seen the development of the gas turbine engine and its application to aircraft propulsion. The gas turbine engine basically consists of a compressor, combustion chamber and turbine. The turbine provides the work input to the compressor. In the case of power production, additional work

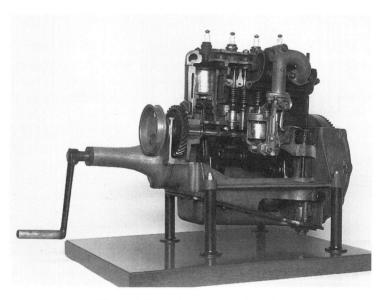

Figure 1.5 An Austin 7 petrol engine.

from the turbine provides external work through a shaft. In the case of a 'jet' engine the turbine provides just enough work to drive the compressor, leaving energy in the gas stream to expand through a nozzle and provide a propulsive force. Gas turbine engines work on the constant pressure cycle, described in Chapter 8.

The first practical engine to work on the basic principles embodied in this cycle was built by an American engineer, George Brayton, in 1873. (In American textbooks on thermodynamics, the constant pressure cycle of the gas turbine engine is called the Brayton cycle in recognition of this fact.) Instead of a rotating compressor and turbine, Brayton used a reciprocating compressor and a separate reciprocating 'expander', as shown in Figure 1.6.

The first rotary gas turbine engine was built in 1903 but it had a low-power output of 8 kW and an extremely low operating efficiency of 3%. Gas turbine technology was hampered by the low efficiencies of rotary compressors until Frank Whittle applied the technology developed in the superchargers of high-performance petrol engines used in aircraft. He developed, first by

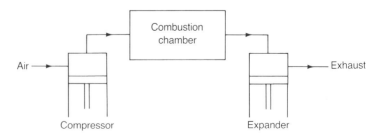

Figure 1.6 Diagram of the Brayton engine.

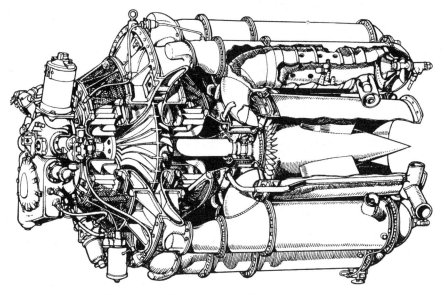

Figure 1.7 Diagram of the Whittle jet engine.

thermodynamic analysis and then by experiment, a gas turbine engine that had sufficiently high component efficiencies for practical jet propulsion. Figure 1.7 shows a diagram of the Whittle jet engine used to power the first jet propelled fighter in Britain.

Subsequent impetus to jet engine development has led to improvements in compressor efficiency, turbine efficiency, combustion chamber technology and the use of high-temperature materials. A typical modern high-performance turbojet engine is shown in Figure 1.8. This improvement of aircraft engine technology has also been transferred to power-producing gas turbine engines. Engines derived from aircraft engines have operating efficiencies of 30%, and above, and are used in the oil and gas industries and as the main propulsion on board some ships.

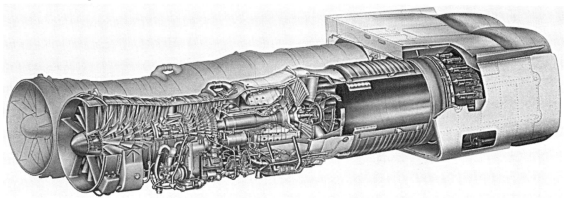

Figure 1.8 Cut away diagram of the Rolls-Royce/SNECMA Olympus 593 turbojet engine.

1.6   SYSTEM OF UNITS

In order to analyse thermodynamic engines or processes it is necessary to have a coherent system of units to define the properties of the fluid under consideration. Throughout this book the International system will be used. This is invariably referred to as the SI system, SI being an abbreviation for Systeme International d'Unites.

Within the SI system there are four basic units corresponding to the four dimensions, from which all other units are derived. The four basic dimensions are length, mass, time and temperature, defined in Table 1.2 with their SI units.

Table 1.2 The four basic dimensions

Dimension	Unit	Symbol
Length	metre	m
Mass	kilogramme	kg
Time	second	s
Temperature	kelvin	K

1.6.1   Derived units

The units of all thermodynamic properties or variables, can be derived from the four basic units.

Velocity is required to define the flow of fluids or the movement of machines such as aircraft. Velocity is the distance travelled during a unit of time. In dimensional terms,

velocity = length/time

with units of

$(m)/(s) = m/s$

Volume defines the amount of space occupied by a quantity of fluid or an object, and has units of (m^3). Specific volume defines the amount of space that is occupied by one unit mass, i.e. 1 kg, of the substance:

specific volume = $(length)^3/mass$

with units of

$(m^3)/(kg) = m^3/kg$

Force has units that are derived from Newton's second law of motion:

force = mass × acceleration

with units of

$(kg) \times (m/s^2) = kg\, m/s^2$

This composite unit for force is appropriately called the newton and is given the symbol (N) where

$N = kg\, m/s^2$

The formal definition of the newton is the force required to accelerate a mass of 1 kg at a constant rate of 1 m/s².

The weight of an object is the force acting on the object as a result of a gravitational field. Again applying Newton's second law of motion:

$$\text{weight} = m \times g \tag{1.1}$$

The value of the gravitational constant depends on position. For example, an object on the surface of the moon would only have one-sixth of the weight it would have on the surface of the earth. Even on earth the value of the gravitational constant varies with height above the surface. For thermodynamic analysis it is adequate to take the gravitational constant as that for normal sea-level conditions:

$$g = 9.81 \text{ m/s}^2$$

Energy can be defined in the same units as work, since both heat and work are forms of energy. From basic mechanics work is given by

$$\text{work} = \text{force} \times \text{distance}$$

with units of

$$(\text{N}) \times (\text{m}) = \text{Nm}$$

This composite unit for energy is called the joule and is given the symbol (J) where

$$\text{J} = \text{Nm}$$

The formal definition of the joule is the energy produced when a force of 1 N acts though a distance of 1 m.

Power is the rate of doing work. In other words, it is the work done during a unit of time:

$$\text{power} = \text{work/time}$$

with units of

$$(\text{J})/(\text{s}) = \text{J/s}$$

This unit for power is called the watt and is given by the symbol (W) where

$$\text{W} = \text{J/s}$$

Since heat and work are both forms of energy, the rate at which heat is transferred has the same unit as power and is also defined by the watt.

Example 1.1

Find the power required to lift a man, with a mass of 75 kg, through a height of 50 m in 10 s. Assume $g = 9.81 \text{ m/s}^2$.

$$\text{weight of man} = 75 \times 9.81 = 735.75 \text{ N}$$
$$\text{work done on man} = \text{weight} \times \text{distance}$$
$$= 735.75 \times 50 = 36787.5 \text{ J}$$
$$\text{power} = \text{work done/time}$$
$$= 36787.5/10 = 3678.75 \text{ W}$$

1.6.2 Unit prefixes

From example 1.1 it will be seen that the amount of energy required to raise the man through the height given is 36787.5 J. The unit joule is a relatively small quantity and for most practical situations involves the use of large numbers. Such large numbers are inconvenient when carrying out an analysis and it is more sensible to use multiples of the unit in order to make the numbers more manageable. Table 1.3 gives a list of the multiplying factors in common use with the SI system.

Table 1.3 Common multiples in the SI system

Multiple	Prefix	Symbol
10^9	giga	G
10^6	mega	M
10^3	kilo	k
10^{-2}	centi	c
10^{-3}	milli	m
10^{-6}	micro	μ

From the prefixes given, energy can be expressed in kilojoules (kJ) where

$$36787.5 \, J \approx 36.8 \, kJ$$

Power can be conveniently expressed in either megawatts (MW) or kilowatts (kW).

For lengths, British Standards advocate the use of either the millimetre (mm) or the metre (m) and engineering drawings are presented in these two units. However, when it comes to volumes, the difference between a cubic millimetre (mm^3) and a cubic metre (m^3) is too great for practical purposes. Therefore, the cubic capacity of a car engine is still quoted in terms of cubic centimetres, (cm^3 or cc). For the flow of liquids, the flow rate is sometimes given in litres, given the symbol 1, where

$$1 \, l = 10^{-3} \, m^3$$

The neat system of prefixes does not apply to time. Although the second (s) is the basic dimension of time in the SI system, longer periods are still expressed in terms of the minute, hour or day. The speed of a motor car is, therefore, expressed in kilometres per hour (km/hr), since it is more convenient than the SI unit of velocity, metres per second (m/s).

1.7 PRESSURE

Pressure is the force exerted by a fluid on a unit area. As such, it is a property of a fluid. The equivalent in a solid is 'stress'. The unit of pressure is defined as

$$\text{pressure} = \text{force/area}$$
$$= \text{force/length}^2$$

with units of

$(N)/(m^2) = N/m^2$

This composite unit for pressure is called the pascal and is given the symbol (Pa) where

$Pa = N/m^2$

The pressure unit Pa is too small to conveniently define the pressure of fluids found in practice so that kilopascal (kPa) or megapascal (MPa) are more commonly used.

The other unit of pressure that may be encountered is the 'bar', where the bar is defined as

$1\ bar = 10^5\ Pa = 100\ kPa$

One advantage of the bar is that it very nearly equals the pressure of one atmosphere. A study of any weather chart will show that atmospheric pressure is not constant, it varies as low- or high-pressure regions cross the country. For this reason a standard atmospheric pressure of 101.325 kPa has been defined as a basis for analysis.

Pressure, as represented in Pa, is absolute pressure which means that it is defined using a datum of zero pressure. So an atmospheric pressure of, say, 100 kPa is an absolute pressure because it is 100 kPa above the datum at which the pressure would be 0 kPa. However, many pressure measuring devices are not calibrated in terms of absolute pressure, but work on the principle of measuring the actual pressure with relation to local atmospheric pressure. The pressure measured using the atmsopheric pressure is called 'gauge' pressure. In order to find the absolute pressure from the gauge pressure it is necessary to add the atmsopheric pressure to the latter. This is illustrated in Figure 1.9 which demonstrates the situation for both positive gauge pressures, above atmospheric, and negative gauge pressures, partial vacuum. The relationship between absolute, gauge and atmospheric pressure can be expressed by the equation

$$P_{abs} = P_{gauge} + P_{atm} \tag{1.2}$$

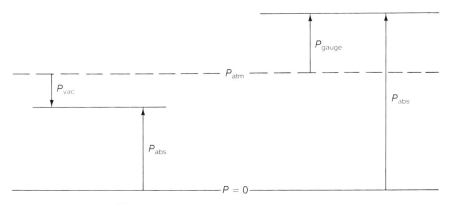

Figure 1.9 Pressure relative to atmospheric.

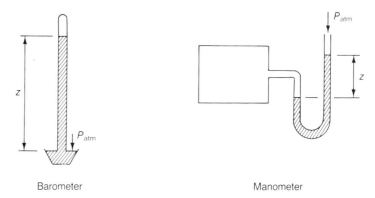

Figure 1.10 Two pressure-measuring devices.

As explained in section 1.3 the realization that the atmosphere had pressure developed in the 17th century. It was found that the atmospheric pressure was equivalent to a height of about 10 m of water. This relationship between pressure and the height of a column of liquid can be used to measure the pressure of a fluid. Figure 1.10 shows two such measuring devices, the first a barometer for measuring atmospheric absolute pressure and the second a manometer for measuring gauge pressure.

A barometer operates by having a vacuum in the end of the closed tube so that the height of liquid is directly related to the atmospheric pressure. The liquid used inside a pressure-measuring device depends on the application. It would be pointless using a liquid in a manometer to measure water pressure if the liquid was soluble in the water.

For a barometer the liquid used is mercury, since it is 13.6 times more dense than water. Typically, standard atmospheric pressure is equal to a height of 760 mm of mercury. Mercury would also be a convenient liquid to use in a manometer for measuring the pressure of water. In the case of gases it is possible to use water as the liquid in a manometer. The relationship between the pressure and the height of liquid z is given by the hydrostatic equation, developed by Massey (1989):

$$P = gz/v \tag{1.3}$$

where v is the specific volume of the liquid.

For greater accuracy of measurement, the height of the liquid z should be as high as possible. This means maximizing the specific volume of the liquid. When measuring the pressure of gases it is possible to enhance the accuracy by using a liquid such as alcohol, with a specific volume greater than water.

Example 1.2
A manometer is used to measure the pressure of a gas in a tank. The liquid used has a specific gravity of 0.0011 m³/kg and the height of the column is 300 mm. If the local atmospheric pressure is 101 kPa, calculate the absolute pressure of the gas. Take $g = 9.81$ m/s².

    height of liquid   = 300 mm = 0.3 m

$$\text{gauge pressure} \quad = \quad \frac{9.81 \times 0.3}{0.0011} = 2675.4 \text{ Pa, ie } 2.68 \text{ kPa}$$

$$\text{absolute pressure} = \quad 2.68 + 101 = 103.68 \text{ kPa}$$

1.8   TEMPERATURE

It is not easy to give a definition of temperature other than it is a measure of the 'hotness' or 'coldness' of a body or fluid. The temperature of a fluid is one of the properties of that fluid, along with pressure and specific volume. It is necessary to have a precise value of temperature in order to define the state of the fluid. If the temperature of a fluid changes then the state of the fluid changes. This is the basis of the thermometer as a temperature-measuring device. A column of liquid inside the thermometer expands or contracts with changes of temperature.

   When a thermometer is brought into contact with a body at a temperature greater than itself, then heat is transferred from that body to the glass of the thermometer and thence to the liquid inside the thermometer. The heat transfer takes place as a result of the temperature difference between the body and the thermometer. As soon as the temperature inside the thermometer is raised to that of the body, there is no further heat transfer and the liquid inside the thermometer reaches a constant level. At this point the liquid inside the thermometer is at the same temperature as the glass of the thermometer which, in turn, is at the same temperature as the outside body.

   This can be expressed formally though what is known as the zeroth law of thermodynamics: 'if two bodies are in thermal equilibrium with a third body, they are in thermal equilibrium with each other'. This may seem to be a statement of the obvious but it serves as the basis for measuring temperatures. This is because any body outside a thermometer and the liquid inside the thermometer will not be in direct contact but they are in thermal equilibrium when the temperature of each is the same.

   It is necessary to have a common temperature scale for measurement. The temperature scale used in the SI system is based on the Celsius scale. This uses two fixed points, the freezing point and the boiling point of water under standard atmospheric conditions, to define 0°C and 100°C respectively. However, this means that for temperatures below 0°C the values become negative. For example, the boiling point of liquid oxygen is –183°C.

   The use of negative temperature values is eliminated in thermodynamics by using an absolute temperature scale. Just as for pressure, where the lowest possible pressure is 0 on the absolute pressure scale, so the lowest possible temperature is 0 on the absolute temperature scale. As explained in section 1.3, the absolute temperature scale is based upon the unit of a kelvin, K, where one kelvin is equal to one degree Celsius.

   The value of absolute zero temperature, 0 K, is –273.15°C. This can be found from experiment using a gas thermometer. If a constant volume of gas is lowered in temperature the gas pressure will also be lowered. This relationship between pressure and temperature can be derived by combining Boyle's law

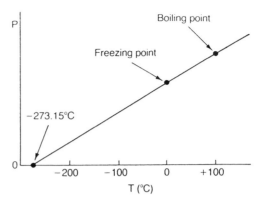

Figure 1.11 Variation of pressure with temperature for a constant volume of gas.

and Charles's law. By plotting the variation of pressure with temperature the results can be extrapolated to predict absolute zero temperature, at which point the pressure will be zero, as shown in Figure 1.11.

A temperature on the Celsius scale can be converted to the absolute kelvin scale by adding 273.15°C to the value of the temperature. However, in engineering analysis the value of absolute zero is considered to be sufficiently accurate if taken as –273°C, so that the conversion between the two scales is given by

$$T\,(\mathrm{K}) = T\,(°\mathrm{C}) + 273 \tag{1.4}$$

Equation (1.4) gives the value of a particular temperature, but what happens in the case of a temperature difference? For example, in a car radiator the water from the engine could be entering at 95°C and the cooling air could be entering at 15°C. Clearly, the difference between the two is 80, but which temperature units apply? The convention is to quote temperature differences in K and the logic behind this becomes obvious when the two original temperatures are converted to absolute temperatures:

$$\text{temperature difference} = (95 + 273) - (15 + 273) = 80\,\mathrm{K}$$

1.9 MOLECULAR UNIT

Avogadro's hypothesis suggests that a constant volume of gas at a given temperature and pressure will contain a fixed number of molecules irrespective of the particular gas being considered. Therefore, a particular volume of hydrogen will have the same number of molecules as the same volume of oxygen. However, oxygen has a higher molecular weight than hydrogen, where the molecular weight is defined as the mass of a molecule of a substance compared to one atom of hydrogen. It follows that the mass of the volume of oxygen is greater than the mass of the volume of hydrogen.

The molecular weight of oxygen can be taken as 16 compared to 2 for hydrogen. A volume that contains 2 kg of hydrogen would contain 16 kg of oxygen. Because of this variation in the molecular mass, it is sometimes necessary to use a unit that is independent of the **actual mass** of the gas. This is particularly true for situations where two, or more, gases are combined together to create a reaction, such as in a combustion process.

This unit is based upon a set number of molecules and is called the 'mole', being defined as the amount of a substance that contains as many molecules as there are atoms in 0.012 kg of carbon-12. In the SI system it is more convenient to use the kilogram mole as the molecular unit. This is given the symbol 'kmol' and, from the definition given above, contains as many molecules as there are atoms in 12 kg of carbon-12. By using this unit the analysis of the behaviour of all gases can be carried out on the same molecular basis.

The actual mass of a gas in kmol depends on the molecular weight of the gas and can be calculated from

$$m = n \times M \tag{1.5}$$

where m is mass of the gas in kg, n the number of kmol and M the molecular weight of the gas. A brief list of useful molecular weights is given in Table 1.4.

Because a given volume of gas contains a constant number of molecules, the kmol can be visualized as being equivalent to a particular volume of gas. However, the volume of a gas varies with both temperature and pressure so that both of these properties have to be specified in order to find the volume in question. This is covered in Chapter 4 during the discussion on the equation of state for perfect gases.

Table 1.4 Molecular weights of some common substances

Substance	Molecular symbol	Molecular weight
Carbon-12	C	12
Hydrogen	H_2	2
Nitrogen	N_2	28
Oxygen	O_2	32

Example 1.3

A kmol of air can be considered to contain 79% of nitrogen and 21% of oxygen by volume. Calculate the mass of air contained.

$$1 \text{ kmol air} = 0.79 \text{ kmol } N_2 + 0.21 \text{ kmol } O_2$$
$$m = (nM)N_2 + (nM)O_2$$
$$= 0.79 \times 28 + 0.21 \times 32$$
$$= 22.12 + 6.72 = 28.84 \text{ kg}$$

Example 1.4

Find the mass of 2 kmol of methane gas. Take the chemical composition as CH_4.

$$\text{molecular weight of } CH_4 = 12 + 2(2) = 16$$
$$m = nM$$
$$= 2 \times 16 = 32 \, \text{kg}$$

SUMMARY

In this chapter the SI system of units has been presented as a basis for the analysis of thermodynamic situations. The key variables and the units used are listed below:

length	m
mass	kg
time	s
temperature	°C
absolute temperature	K (K = °C + 273)
velocity	m/s
specific volume	m^3/kg
force	N
energy	J
power	W
pressure	Pa
molecular mass	kmol

PROBLEMS

1. Express the watt, the unit of power, in terms of the basic SI dimensions of length, mass and time.
2. Express a time period of $2\frac{1}{2}$ hours in the basic time dimension of a second, using a suitable prefix.
3. The pressure of water in a tank is measured by means of a mercury-filled manometer. If the height of the mercury is 500 mm and the atmospheric pressure is 100 kPa, calculate the absolute pressure of the water. Assume the specific volume of mercury to be $7.35 \times 10^{-5} \, m^3/kg$. Take g = 9.81 m/s².
4. The pressure of air inside a pipe is being measured by means of a water manometer. Find the absolute pressure of the air if the height of water in the manometer is 400 mm on a day when the barometer reads 750 mm of mercury. Take the specific volume of water as 0.001 m³/kg, the specific volume of mercury as $7.35 \times 10^{-5} \, m^3/kg$ and g = 9.81 m/s².
5. The temperature of a human body is about 37°C. Quote this temperature in absolute temperature units.
6. A mixture of gases consists of 1 kmol of carbon dioxide to 4 kmol of nitrogen. Find the mass of 1 kmol of the mixture.

Modelling $\boxed{2}$

2.1 AIMS

- To introduce the idea of using different types of model in problem solving.
- To define the types of model used in visualizing thermodynamic situations.
- To introduce a thermodynamic system as a conceptual model of a device or situation.
- To identify the thermodynamic processes performed within a device.
- To introduce the concept of reversible or irreversible processes.
- To explain the conservation principles used in the mathematical modelling of thermodynamic processes and cycles.

2.2 PROBLEM SOLVING

The purpose of studying thermodynamics is to be able to solve problems. To do this the student must have a good understanding of the physical behaviour of the fluid used in a thermodynamic device or situation. Typical devices that will be considered in this and later chapters are turbines, compressors, boilers, and pistons in cylinders. In order to analyse such devices it is also necessary to be able to model the situation.

A model is a means of representing the real device or situation. The most familiar type of model is probably the 'scale' model. For example, a car manufacturer would test a scale model of a proposed design in a wind tunnel to check on the drag characteristics, before going to the expense of building a full-sized prototype. Such scale models are widely used in engineering but are outside the scope of this book as model engines often provide little information of use in the analysis of their full-sized counterparts. This is due to what is called the 'scale effect'. The friction and thermal losses for a model are proportionally very much greater than for the full-sized engine. The surface area of a component, as a ratio of its contained volume, increases as the component gets smaller resulting in proportionally greater thermal losses.

From an opposite point of view, thermal losses are reduced as the size is increased which is why some of the larger mammals, such as elephants, tend to 'overheat'.

The types of model that are used to define and analyse thermodynamic situations are broadly classified as:

1. **the conceptual model** – used to visualize the device and processes that take place within it;
2. **the mathematical model** – used to derive mathematical equations in order to define the thermodynamic behaviour of the working fluid under the relevant conditions.

Using such models it is possible to outline an ordered approach to the solving of thermodynamic problems. Although not all problems are the same, or are solved in exactly the same way, it is necessary to have a clear idea of how to tackle problems. The following step-by-step approach provides a basis for solving problems.

1. **Understand the problem**. An obvious statement but one that needs emphasizing. Understanding means having a clear picture of what the problem entails and this, in turn, means creating the necessary conceptual model.
2. **Define the problem**. In thermodynamic problems it is possible to model the device or situation in terms of a thermodynamic system. The system may be either closed or open, depending on the situation.
3. **Draw a representative diagram of the processes**. Process diagrams are an essential component in the solving of thermodynamic problems. By definition, such diagrams are conceptual models of the fluid behaviour within a thermodynamic system and define the behaviour in terms of the fluid properties.
4. **Build a mathematical model of the problem**. Application of the basic equations of mass, energy and momentum in order to define the processes taking place in a thermodynamic system.
5. **Make assumptions**. Simplify the general equations to make them more applicable to a particular problem by ignoring the terms that have negligible effect. For example, if the velocity is very low the kinetic energy of a mass of fluid can be considered insignificant.
6. **Analyse the problem**. Use the simplified equations in a re-arranged form to determine the unknown quantities within the problem.

Problem solving becomes easier to visualize through the solution of actual problems, an illustration of the old adage that practice makes perfect.

Nevertheless, the step-by-step approach to problem solving outlined above serves two functions. First, it illustrates that problem solving is an **ordered activity** and, secondly, steps 1–4 show that models are the basis of problem solving. It therefore serves as an introduction to the following discussion on the particular types of model used in thermodynamic problems.

2.3 UNDERSTANDING THE PROBLEM

There are three types of conceptual model used to gain a clear picture of the problem.

2.3.1 Draw a sketch

It is a truism that one cannot solve what one cannot see. It pays, therefore, to sketch the physical situation defined within the problem. Figure 2.1 shows a typical steam plant incorporating an oil-fired boiler. The sketch shows both the oil tank and the cylindrical boiler. From the boiler the steam is fed into a reciprocating steam engine that produces shaft work. The steam leaves the engine at a lower pressure than it entered and is condensed into water in a condenser. The water is then fed back into the boiler by means of a pump.

Figure 2.1 is a conceptual model of a steam plant. It shows the basic features whilst at the same time leaving out an amount of fine detail. The sketch shows a pictorial view of the plant and is therefore easy to comprehend, even for those with limited knowledge of such plants. These are all positive merits but the real question that has to be asked is 'Does Figure 2.1 help us to analyse the steam plant?'. The answer is a rather hesitant 'Yes'.

Yes, because it helps to visualize the plant and therefore presents a clearer picture of the problem than no sketch at all. Hesitant, because it is simply a pictorial representation, is time consuming to draw and gives no technical data concerning the plant. There is no certainty that the various components are in the relative positions shown. The final arrangement might simply be an artistic impression to fit within a particular diagram.

Figure 2.2 shows the same steam plant but in an entirely different manner. This diagram shows a schematic arrangement of the plant, highlighting the important working components. In addition, Figure 2.2 has the advantages over Figure 2.1 of being simpler and therefore quicker to sketch, and of clearly illustrating the thermodynamic situation in greater detail.

The engine is portrayed as an expander since the steam enters at high pressure and is expanded to a lower pressure in order to produce work. The expansion can take place either within a reciprocating engine or within a rotating turbine. The two devices work differently but, in principle, if the inlet and outlet conditions of the steam are the same, then ideally the work produced by either will be the same. It is not necessary to distinguish which particular type

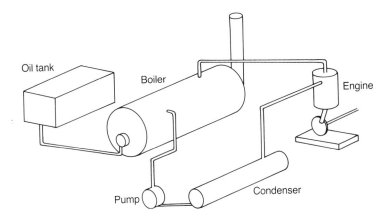

Figure 2.1 Sketch of steam plant.

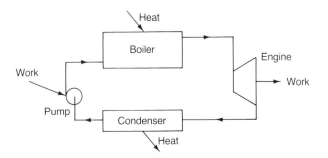

Figure 2.2 Schematic diagram of a steam plant.

of expander is used and the indication of an expander on the diagram is sufficient.

From Figure 2.2 it is quite obvious that the working fluid circulates around the plant in a continuous manner. The pipe work is shown as lines joining the components, whilst the arrows indicate the direction of flow. This information is not so apparent from the pictorial representation in Figure 2.1.

Finally, it is possible to indicate how heat or work is transferred in or out of the different components of the plant. In Figure 2.1 it was shown that the boiler had a supply of oil from a tank to a burner. This gives too much intricate detail. From the point of view of solving a thermodynamic problem it is sufficient to indicate that heat is transferred into the boiler. Figure 2.2 is therefore a far more relevant way of drawing a sketch than Figure 2.1 since it gives a clearer picture of the problem being considered; such sketches are used throughout this book as a basis for problem solving.

2.3.2 A different viewpoint

Figures 2.1 and 2.2 show a steam plant that is stationary. It is possible to visualize the plant and analyse it because it is stationary. Now suppose that the steam plant was used on board a ship to propel it along. How can this situation be visualized? Can it be achieved, for example, by considering the ship as it moves towards us at a velocity of, say, 4 m/s and then as it recedes to the horizon in the opposite direction at 4 m/s? The answer is no. It is conceptually much easier to consider that the person observing the steam plant is aboard the ship and therefore, relative to that person, the steam plant is stationary.

It is, therefore, necessary to consider the viewpoint from which the observer is looking at any thermodynamic situation. With most problems the device being analysed will be stationary and this will be obvious without needing to state it.

However, consider the situation shown in Figure 2.3(a). An aircraft is flying above the ground and is propelled by an engine. To the observer on the ground the aircraft is moving forward with a velocity of, say, 250 m/s. In principle the problem is the same as the steam plant on the ship, but because the ship is only moving at 4 m/s whereas the latter is travelling at 250 m/s, the aircraft problem is more difficult to visualize.

To the observer on the ground in Figure 2.3(a) the ground and air are stationary, whereas the aircraft is moving at a velocity of 250 m/s. If the viewpoint is now changed so that the observer is on the aircraft, the aircraft now appears to

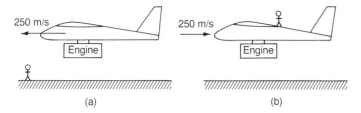

Figure 2.3 Relative motion of an aircraft.

be stationary and it is the ground that is moving below at 250 m/s. More importantly, the air is moving towards the aircraft and the engine at 250 m/s.

From the point of view of analysing the performance of an aircraft engine, the situation shown in Figure 2.3(b) is easier to picture and to analyse than the real situation in Figure 2.3(a). This type of visualization is used in Chapter 9 to allow the performance of aircraft engines to be analysed.

2.3.3 By analogy

An analogy is a way of studying a thermodynamic situation by considering an equivalent situation in a different field of study. As such, it is used within general engineering as a powerful means of problem solving. It is sometimes referred to as the art of making the strange familiar.

Several examples of analogy can be derived from the similarity between thermal devices and electrical devices. Thermal insulation is analogous to electrical insulation. Alternatively, a thermal resistance is analogous to an electrical resistance. So, in Chapter 12, the heat transfer across a combination of walls and fluids can be solved by considering the overall situation as a network of electrical resistances with each electrical resistance equivalent to a thermal resistance in the composite situation. This analogy allows the situation to be viewed in a simpler manner and allows quite complex situations to be analysed more easily.

A valve in a pipeline is analogous to an electrical switch in a wire. In fact, this type of analogy has already been introduced as the schematic diagram of a steam plant in Figure 2.2. The plant is represented in a similar form to an electrical circuit.

In Chapter 6 there is a discussion of the application of the second law of thermodynamics. This is vitally important to the understanding of the functioning of a heat engine. The original idea was developed by Carnot who used the known characteristics of a water wheel as analogous to a heat engine.

So, although analogies have only limited application within this book, they nevertheless represent an important type of conceptual model.

2.4 THERMODYNAMIC SYSTEM

A thermodynamic system is the next stage in the conceptual modelling of a device or situation. However, before defining a thermodynamic system in detail it is necessary to discuss the use of the word 'system' in more general terms.

A system is a term so widely used in engineering that its use can cause a certain amount of confusion. An engineering system is a collection of interconnected parts or components. In a single-cylinder petrol engine the engineering system comprises the cylinder, piston, connecting rod and crankshaft together with associated parts such as bearings and valves. The space contained within the cylinder and piston assembly is the region in which the thermodynamic changes take place. In order to analyse the thermodynamic behaviour it is necessary to define this region by means of a thermodynamic system, a thermodynamic system being a collection of matter in the form of a working fluid. There are two types of thermodynamic system, a closed system and an open system.

2.4.1 Closed system

A closed system is one where there is no flow of fluid into or out of the region under consideration. It therefore follows that a closed system has a fixed quantity, mass, of fluid inside.

A closed system is shown in Figure 2.4. This diagram shows the main parts of a thermodynamic system, the closed system itself, the 'boundary' containing it and the region outside the boundary called the 'surroundings'. By surroundings we mean any region outside the system. In practice it would be difficult to visualize a system having a significant influence on the whole earth. A petrol engine can be assumed to have no effect on the atmospheric conditions or content at, say, a distance of 10 km away. The surroundings are therefore considered to be those in the locality of the system.

The boundary of a closed system fulfils two functions. It can be considered to be a boundary between the system and the surroundings, and retains the fixed quantity of fluid within the system. It also provides a necessary conceptual model for problem solving by defining the particular region under consideration. By defining the boundary the closed system is defined.

Figure 2.5 shows a cylinder and piston assembly containing a gas. The gas can be assumed to be at a higher pressure than the surroundings, causing the piston to move. If we take the region encompassed by the cylinder walls and the top of the piston as the closed system, then the boundary is defined by A. As the piston moves the boundary will have to change shape to accommodate the movement. A boundary is not rigid and fixed but, providing that there is a given mass of fluid within the system, the boundary can be movable.

By comparison, boundary B defines the same situation but now the system encompasses the gas within the cylinder, the cylinder and piston assembly, and part of the region around – a different way of looking at the same situation, simply by changing the boundary. By making boundary B large enough the region surrounding the cylinder would not be influenced by the movement of the piston whereas, for boundary A, the movement is very significant as the moving piston would do work on the surroundings.

The question must obviously be raised as to which is the correct boundary, A or B? The answer is both since both define a closed system! It is the analysis required, and the information that is derived from that analysis, that determines which is the most relevant.

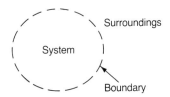

Figure 2.4 A closed system.

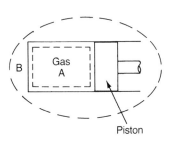

Figure 2.5 Cylinder and piston assembly.

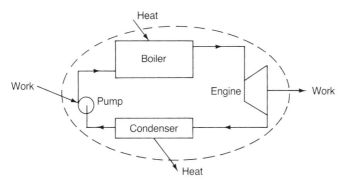

Figure 2.6 Steam plant in a closed system.

By considering boundary A the model can be used to analyse the work done by the moving piston. If this was the aim, then clearly boundary A is correct and boundary B is incorrect, in the context of the information to be found.

The movement of the piston and the resulting work done on the surroundings means that work can cross the boundary of a closed system. In fact, both heat and work can cross the boundary as both are forms of energy and are not properties of the system. Both heat and work are discussed in greater detail in the next chapter, but it is helpful to consider one more example of a closed system, namely the steam plant defined in Figure 2.2.

As the working fluid in the steam plant is a fixed quantity, then by definition the plant is a closed system and the boundary can be defined as shown in Figure 2.6. Since heat and work can cross the boundary the diagram also defines the heat transfer into the boiler, the work transfer from the expander, the heat transfer from the condenser and the work transfer into the feed pump.

2.4.2 Open system

The steam plant can be modelled as a closed system but what about the individual components that go to make up the system. In the boiler, for example, there is a flow of water in and a flow of steam out. The amount of fluid within the boiler is not fixed, because of this flow. The boiler is therefore not a closed system but is a typical example of an open system. By defintiion, an open system allows a flow of fluid across the boundary.

Just as a closed system is defined by the boundary so an open system is also defined. Figure 2.7 shows a boiler as an open system. There is heat transfer across the boundary together with the mass flow of water entering and steam

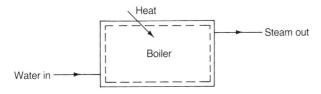

Figure 2.7 A boiler as an open system.

leaving. However, unlike the boundary of a closed system, the boundary of an open system is not movable but is fixed in space. Therefore, an open system can be defined as a fixed volume of space. Because of this an open system is sometimes referred to as a 'control volume' and this term will be found in textbooks on fluid mechanics such as Massey (1989) and occasionally in textbooks on thermodynamics. However, within this book the term 'open system' will be used throughout.

The four components in the steam plant defined in Figure 2.6 – boiler, expander, condenser and feed pump – can be analysed individually by modelling each as an open system.

2.5 THERMODYNAMIC PROPERTIES

The state of a system is determined by the values of the properties of the fluid within it. Two such properties were discussed in the previous chapter, namely pressure and temperature. These are particularly relevant as they can be measured directly and can, therefore, be visualized, if not as a concept, at least as a position on the graduated scale of a pressure gauge or thermometer.

Other properties that can be measured directly are the volume, mass and velocity of the fluid within the system. In the case of the pressure, temperature and velocity, the property can change with time so that any particular value must be defined at a particular instant. It follows that the state of a system also changes with time as the system undergoes a process.

However, the velocity is a different type of property to either pressure or temperature. A fluid can have a particular temperature irrespective of whether it has a velocity or not. Velocity is considered to be a mechanical property because it defines the kinetic energy within the system. Another mechanical property is the height of a system within a gravitational field, since this defines the potential energy of the system.

Pressure and temperature are two of the thermodynamic properties and, as such, are independent of the size of the system. A cup of tea at 90°C has exactly the same temperature as a pot of tea at 90°C. Both pressure and temperature are independent of the mass of the fluid contained within a system.

On the other hand, volume does depend on the mass of the fluid. A mass of 2 kg will have twice the volume of a mass of 1 kg, providing the fluid is at the same pressure and temperature. A particular mass has its own unique values of those thermodynamic properties that depend on the quantity of fluid. Volume is one such thermodynamic property. Other thermodynamic properties are internal energy, enthalpy and entropy, which are all defined in the next chapter.

To overcome the inconvenience of relating these properties to a particular mass of fluid, the values can be quoted in 'specific' terms. Namely, the value of the property with relation to a unit mass of the fluid. In the case of specific volume, the units are m^3/kg. Simlarly, values of:

specific internal energy
specific enthalpy
specific entropy

are all quoted with respect to a kg.

A study of thermodynamics is based upon the assumption that the state of a system can be defined in terms of just two thermodynamic properties, providing those two are independent of each other. For example, in the case of a gas, pressure and temperature are independent of each other. Defining a particular pressure and a particular temperature, defines the specific volume which, in turn, defines the other specific properties.

On the other hand, a boiling liquid has a temperature that depends on the pressure. In this case the pressure and temperature are not independent, so that the state of the fluid would need to be defined in terms of either pressure or temperature, and one other independent thermodynamic property.

2.6   THERMODYNAMIC PROCESSES

A thermodynamic process can be defined in terms of the fluid properties within a system. The properties define the state of the fluid but, for the properties to be measured or calculated, the fluid within the system must be in a condition of 'equilibrium'.

2.6.1   Thermodynamic equilibrium

Equilibrium implies a stable condition in which there is no inherent imbalance. In the case of mechanical equilibrium all the forces on a component are completely balanced. This is well accepted and is embodied in Newton's Third law of motion – 'To every force there is an equal and opposite reaction'.

So it is with thermodynamics. It would be extremely difficult to analyse a situation in which the fluid was unstable and could change its state in an unpredictable manner. Therefore, thermodynamic processes are defined on the assumption that the fluid within the system is in thermodynamic equilibrium.

This means that the fluid must comply with three types of equilibrium. The first of these is thermal equilibrium, in which the temperature is constant. Figure 2.8 shows a closed system. In Figure 2.8(a) the fluid is not in thermal equilibrium as there are temperature differences. These temperature differences could give rise to heat transfer within the fluid and to motion of the fluid within the system. In Figure 2.8(b) the fluid has achieved equilibrium conditions as the temperature is constant throughout the system.

Similarly, the fluid must have constant pressure throughout the system otherwise the forces within the fluid would be out of balance and the system would not have mechanical equilibrium. Finally, the fluid must have chemical equilibrium so that its chemical composition does not change with time. Therefore, equilibrium is an essential assumption when modelling thermodynamic situations.

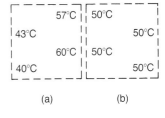

Figure 2.8 A closed system.

2.6.2   Processes

Any changes that a system undergoes when the fluid inside moves from one equilibrium state to a new equilibrium state is called a 'process'. For any pure fluid it is sufficient to define the state of the fluid by means of two independent

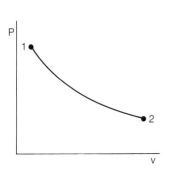

Figure 2.9 A thermodynamic process.

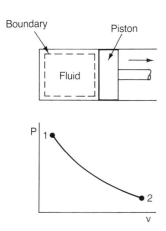

Figure 2.10 Expansion process.

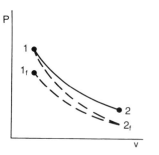

Figure 2.11 Expansion and compression process.

properties. This is extremely useful when visualizing thermodynamic processes, as each process can be defined at each stage by just two properties and a graph of the process drawn using the two properties as the ordinates.

Figure 2.9 shows the graph of a fluid being expanded from pressure P_1 and specific volume v_1, down to a lower pressure P_2. As the fluid expands the specific volume increases.

Since pressure and specific volume are independent of each other, it follows that Figure 2.9 contains all the information required to define the process. Knowing the characteristics of the fluid means that other properties, such as temperature, can be found from the values of pressure and specific volume. Similarly, the other properties of the fluid can be found.

A process diagram of the type shown in Figure 2.9 is an effective way of visualizing the behaviour of a system. Two such process diagrams that form the basis of analysing thermodynamic situations are the pressure–specific volume and the temperature–specific entropy diagrams. Examples of these will be found in the following chapters of this book.

However, following the previous discussion on equilibrium, the whole process must be in equilibrium for it to be capable of being defined and analysed. This not only means the two end states at points 1 and 2, but at all intermediate states as well.

Figure 2.10 shows the expansion process shown in Figure 2.9 in greater detail. The fluid is contained within a cylinder having a sliding piston. As the piston moves to the right the fluid inside the closed system expands. If the movement of the piston is relatively slow then the process can be assumed to be in or near equilibrium at all times.

On the other hand, if the movement is extremely fast there will be a delay while the molecules of the fluid adjacent to the piston re-arrange themselves in order to follow the movement of the piston. This would cause a variation of pressure in the fluid and the process would not be in equilibrium.

In practice this does not happen as the movement of a piston is always going to be slower than the velocity at which molecules react to the change. What does happen is that the fluid in motion may create turbulence as it expands to fill the volume contained by the cylinder and piston. This turbulence is the result of local variations of the velocity of the fluid. Fortunately, the magnitude of the velocities are small and can be considered to have negligible effect on the process. Therefore, the process can be assumed to be in equilibrium.

2.6.3 Reversibility

When a process is reversed such that the fluid achieves the original state, then it is said to be 'reversible'. The concept of reversibility is an extremely useful one in thermodynamics and will also be discussed in Chapter 6 which covers the Second law. The present discussion forms a brief introduction within the context of processes.

The cylinder and piston shown in Figure 2.10 result in an expansion process. If, instead of the piston just moving to the right, it subsequently moves to the left, the gas first expands and is then compressed. Figure 2.11 shows the possible processes that can result.

If the gas expands from 1 and the piston moves freely, without friction, the process is ideal and ends at point 2. Reversing the movement of the piston causes the gas to be compressed back to point 1. Therefore, the original expansion process is reversible.

On the other hand, if there is friction between the piston and cylinder, there will be a greater fall in pressure during the expansion and the process will follow the curve from 1 to 2_f, the subscript 'f' in this case referring to the resulting end state with friction. Compressing the gas back to the original volume results in end state l_f which is away from point 1. The expansion from 1 to 2_f is therefore not reversible. Such a process is called 'irreversible' and the irreversibility is attributable to the friction of the piston.

Generally, in thermodynamics, processes in cylinders with sliding pistons are assumed to be reversible as the friction can be ignored. In rotating devices such as compressors or turbines, the flow of fluid through the blades is subject to friction and turbulence so that the process must be assumed to be irreversible.

Friction is not the only cause of irreversibility. Some processes or devices are inherently irreversible. For example, an electric current flowing through a heating coil causes heat transfer to the surroundings. Reversing the process by transferring heat from the surroundings to the coil does not generate an electric current and so the situation is irreversible. Other irreversible processes will be discussed in later chapters.

2.6.4 Cycles

When a fluid undergoes a series of different processes and returns to its initial state it is said to have undergone a 'cycle' or, more precisely, a thermodynamic cycle. The steam plant defined in Figure 2.2 operates on a continuous cycle. The steam plant is redefined in Figure 2.12(a) to show the state points either side of each component. The process in each component can then be defined on a pressure–specific volume diagram as shown in Figure 2.12(b).

Between state points 1 and 2 the fluid enters the boiler as water, having a low specific volume. The process in the boiler is at constant pressure with increase in specific volume as the water vaporizes. Expansion takes place between points 2 and 3 down to a lower pressure, the vapour is then condensed back to

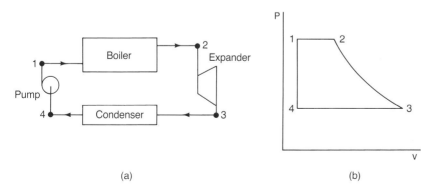

Figure 2.12 Cycle of a steam plant

water between 3 and 4. The water leaving the condenser is pumped back to the boiler pressure, it being assumed that the increase in pressure has negligible effect on the specific volume of the water.

Figures 2.12(a) and (b) represent two forms of conceptual model used to describe a steam plant. The merits of the schematic diagram shown in Figure 2.12(a) have already been discussed earlier. The cycle shown in Figure 2.12(b) allows a clearer insight into the thermodynamic processes taking place within the plant. In Chapter 7 it will be found that an even more useful process diagram for describing the cycle, is the temperature–specific entropy diagram.

2.7   MATHEMATICAL MODELLING

In order to analyse any thermodynamic situation or process it is necessary to derive and apply equations. These equations are mathematical models of the real situation and are based on the knowledge that quantities are conserved. In thermodynamics the three quantities that are conserved are:

mass
energy
momentum

Just as equilibrium requires that a system is in balance, so the conservation principle states that the quantities of mass, energy and momentum are in balance. This means that mass, energy and momentum cannot be created or destroyed.

2.7.1   Conservation of mass

Conservation of mass for a system can be expressed in the form:

Mass entering   –   Mass leaving   =   Gain (or loss) of mass in
    system   system   system

Clearly in the case of a closed system there is no flow across the boundary and the mass inside the system is fixed, so the mass is conserved by definition.

For an open system there is flow across the boundary and mass can both enter and leave, so the conservation of mass does apply. If the mass flow rate entering an open system exactly equals the mass flow rate leaving, and there is no change of mass flow rate with time, the flow is said to be 'steady'.

As an example of a steady-flow open system, consider the model of a windmill shown in Figure 2.13. The rotating blades can be considered to form a disc and Figure 2.13 shows a side view of the disc. This may seem rather removed from the study of thermodynamics, but the example is useful as it embraces conservation of all three quantities – mass, energy and momentum. Also, by a stretch of the imagination, it can be considered to be a thermodynamic device as the energy required to create the wind is a result of heat being transferred from the sun to the atmosphere.

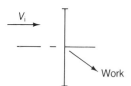

Figure 2.13 Model of a windmill.

Assuming the windmill to operate under steady conditions, the air enters the rotor with constant inlet velocity V_i. As depicted in the figure, the rotor is very thin compared to its diameter and there is no storage of air within. So, to comply with the principle of conservation of mass,

mass flow rate in = mass flow rate out

However, since there will be some transfer of energy from the air flow to the rotor, in order to generate work, the velocity of the air leaving V_o must be lower than the inlet velocity.

The mass flow rate is the volume flow rate, velocity flow area, divided by the specific volume of the air:

mass flow rate $= \dot{m} = VA/v$

It can be assumed that the air at both inlet and outlet are at the same atmospheric pressure and temperature, the specific volume of air is the same and

$$V_i A_i = V_o A_o$$

Since

$$V_o < V_i$$

it follows that

$$A_o > A_i$$

The rotor does not change in cross-sectional area between the inlet and the outlet, so the actual air flow must change in geometry. The resulting envelope of the flow across the rotor forms the boundary of the windmill as an open system, as shown in Figure 2.14.

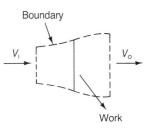

Figure 2.14 A windmill defined as an open system.

2.7.2 Conservation of energy

The conservation of energy for a system can be expressed in the form:

energy entering system   –   energy leaving system   =   gain (or loss) of energy in system

In analysing a system there are several forms of energy that apply in thermodynamic situations. The two most easily recognized from a knowledge of general mechanics, are 'kinetic energy' and 'potential energy'.

Kinetic energy is due to motion, and for a body with mass m the magnitude is given by

$$KE = \tfrac{1}{2}mV^2$$

where V is the velocity of the body.

Potential energy is due to the height of a body in a gravitational field:

$$PE = mgz$$

where z is the height of the body.

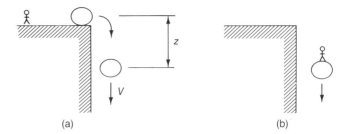

Figure 2.15 Two ways of viewing a falling rock.

The magnitude of both the velocity and the height and, therefore, the magnitude of the energies, depend on the position of the observer. As an illustration of this, consider the falling rock shown in Figure 2.15. A rock falls from the top of a cliff. At the top of the cliff the rock is stationary. As it falls the rock accelerates in the gravitational field to a velocity, V.

To the observer at the top of the cliff, Figure 2.15(a), the total energy of the rock is the same but some of the potential energy has been lost with a resulting increase in kinetic energy. These changes can be equated by

$$mgz = \tfrac{1}{2}mV^2$$

ignoring such factors as the air resistance of the rock during its fall.

However, by considering a different observer standing on the rock during its fall, as shown in Figure 2.15(b), the changes in potential energy and kinetic energy are totally different. Because this second observer is moving with the rock, then, relative to this observer the rock has neither changed height nor changed velocity. For this second observer the total energy of the rock is the same as for the first observer but now there is zero change in potential and kinetic energy.

It is therefore essential to clearly define the position of an observer when establishing the magnitude of potential and kinetic energy.

In the case of a thermodynamic system the position of an observer must be relevant to the particular situation. There is no point in considering the viewpoint of an observer on one of the moons of Jupiter if the thermodynamic system is on the surface of the Earth. Also, by considering the observer in one position, that position is chosen to give the most useful conceptual model as well as providing the data for subsequent analysis. The observer on the top of the cliff provides a greater insight into the energy changes for the falling rock.

An open system is a fixed volume in space and can be considered to be stationary with respect to an observer. A closed system can have a moving boundary but some part of the boundary will remain fixed relative to an observer outside the system.

The cylinder and piston shown in Figure 2.16 form a closed system but the cylinder can be assumed to be stationary relative to an outside observer. The observer can then assess the relative movement of the piston to find the magnitude of the potential energy and kinetic energy of the fluid within the system.

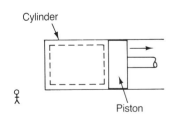

Figure 2.16 Cylinder and piston.

The conservation of energy for a system forms the basis of the first law of thermodynamics, discussed in the next chapter. There is no point in pre-empting that discussion by considering the conservation of energy in greater depth at this stage. However, it is useful to consider the windmill example to illustrate the analysis of energy changes for a system.

Example 2.1

Air enters a windmill with a velocity of 10 m/s and leaves with a velocity of 8 m/s. Calculate the specific work done by the windmill. Note that the specific work done is the work done per unit mass of working fluid, w.

Conceptual model

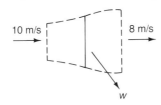

Mathematical model – assume:

1. no heat transfer across the boundary;
2. zero change of potential energy;
3. zero gain of energy by system.

Then,

$$KE_{in} - KE_{out} - w = 0$$

and

$$w = \tfrac{1}{2}mV_i^2 - \tfrac{1}{2}mV_o^2$$

For unit mass of air,

$$w = \tfrac{1}{2}(10)^2 - \tfrac{1}{2}(8)^2 = 18 \text{ J/kg}$$

Example 2.2

If the windmill in example 2.1 has a rotor diameter of 10 m, calculate the power generated. Assume the specific volume of air to be 0.82 m^3/kg.

Conceptual model – as for example 2.1.

Mathematical model:

$$\text{power} = \text{rate of change of work done}$$
$$= w \times \dot{m}$$

Checking units,

$$\frac{\text{J}}{\text{kg}} \times \frac{\text{kg}}{\text{s}} = \frac{\text{J}}{\text{s}} = W$$

For mass flow rate,

$$m^{\cdot} = \left(\frac{10+8}{2}\right) \times \pi \, 5^2/0.82$$

$$= 862 \text{ kg/s}$$

Therefore,

$$\text{power} = 18 \times 862 = 15\,516 \text{ W i.e. } 15.5 \text{ kW}$$

2.7.3 Conservation of momentum

The conservation of momentum for a system can be analysed by applying Newton's second law. This is generally presented in the form

$$F = ma$$

where an external force F applied to a body of mass m results in the body having an acceleration a. It is possible to apply this to an open system as a steady flow of fluid results in an external force. A typical example of such an open system is the jet engine in which air is taken in at flight velocity and leaves with a much higher jet velocity, producing thrust.

Even though the jet engine can be considered to be a steady flow open system there is a resultant thrust since there is a rate of change of momentum across the engine. This can be illustrated by restating Newton's second law in the form

$$F = m \frac{dV}{dt}$$

since the mass is constant, this can be expressed as

$$F = \frac{d(mV)}{dt} = V \frac{dm}{dt} + m \frac{dV}{dt}$$

However, dV/dt is zero, so that

$$F = V \frac{dm}{dt} = m^{\cdot} V$$

For an open system the conservation of momentum can be expressed as

$$\text{External forces} \quad + \quad \begin{array}{c} \text{Rate of change} \\ \text{of momentum in} \end{array} \quad - \quad \begin{array}{c} \text{Rate of change} \\ \text{of momentum out} \end{array} \quad = 0$$

The application of the conservation of momentum can be best illustrated by applying it to the windmill example. Due to the change of velocity across the open system there will be a resultant force applied to the rotor.

Example 2.3
Calculate the resultant force on a 10 m diameter windmill rotor with an air velocity entering of 10 m/s and leaving of 8 m/s.

Conceptual model

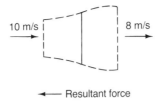

10 m/s → ⟶ 8 m/s →

◄—— Resultant force

Mathematical model

$$F + m\dot{}V_i - m\dot{}V_o = 0$$
$$F = m\dot{}(V_o - V_i)$$

From example 2.2, $m\dot{} = 862\ \text{kg/s}$ and so

$$F = 862\ (8 - 10) = -1724\ \text{N}$$

Note – the negative sign indicates that the resultant force F is in the opposite direction to that assumed in the conceptual model.

2.7.4 Making assumptions

Making assumptions is a very important part of mathematical modelling as the choice of suitable assumptions greatly simplifies the equations used in thermodynamics. There are no written rules regarding the assumptions that can be made as each problem may be subject to different assumptions. The main requirement is to ignore those aspects of the problem that are likely to have negligible influence on the answer.

Taking example 2.1, there were three assumptions that had to be made in order to apply the conservation of energy principle. However, there were several more that were assumed but not stated.

1. The working fluid, air, was completely homogeneous.
2. The rotor was made up of rotating blades that formed a complete disc.
3. The air flowed axially through the rotor and there was no radial loss outwards from the tips of the blades.
4. There was no friction of the rotor on its central axis.
5. There was no air resistance for the rotating blades.

Assumptions 1 and 2 above can be considered to be general since to regard air as anything other than homogeneous and the rotor as anything other than a disc would make the mathematical modelling extremely complex.

Assumptions 3–5 allow the problem to be idealized. The radial loss of air and friction of the rotor, would cause a reduction of the power output. These assumptions are made in the hope that the losses have only a small effect on the power output.

Clearly, in any thermodynamic problem there are going to be basic assumptions made. Not all the assumptions will be stated, since this would make each problem very cumbersome, but the reader should be aware of their existence.

SUMMARY

In this chapter the basic concepts of modelling thermodynamic situations have been discussed. The key terms that have been introduced are:

conceptual model
mathematical model
thermodynamic system
closed system
open system
thermodynamic process
equilibrium
reversibility
thermodynamic cycle
conservation of mass
conservation of energy
conservation of momentum

PROBLEMS

1. Waste energy from a car engine is rejected to the air by means of water circulating through a radiator. Should the radiator be considered as a closed or open system?

2. A potato is being cooked in an open saucepan of boiling water. Considering the potato and the saucepan as separate systems, define the type of system for each.

3. A hand-held hair dryer can be modelled as a tube containing a fan and heating element. What type of system can be used to describe it and define the forms of energy crossing the boundary?

4. In a reciprocating model steam engine the steam flows into the cylinder and moves a piston throughout the whole of the power stroke. Define the boundary that would allow this situation to be considered as a closed system.

5. The model steam engine in problem 4 is connected to a model boiler by a pipe. The steam from the engine is exhausted to the atmosphere and the engine runs until the boiler is empty. Can the boiler and engine be considered as a closed system?

6. Relative to an observer on the ground, find the potential energy and kinetic energy of an aircraft with a mass of 100 tonnes, flying at an altitude of 5000 m with a speed of 200 m/s. Take 1 tonne = 10^3 kg.

7. A petrol tanker has a mass of 5 tonnes when empty. The tank is cylindrical with a length of 4 m and diameter of 2 m. Find the kinetic energy when full if the tanker is travelling at 15 m/s. Assume petrol has a specific volume of 0.0012 m^3/kg.

8. Water flows through the nozzle of a fireman's hose at the rate of 75 kg/s. Find the outlet velocity if the outlet diameter is 70 mm. Also calculate the height the water would reach if the nozzle was pointed vertically upwards. Take the specific volume of water as 0.001 m^3/kg.

9. Air flows though a propeller at the rate of 100 kg/s. If the air enters the propeller at a velocity of 100 m/s and leaves with a velocity of 180 m/s, calculate the thrust developed. Assuming that there are no losses, calculate the power required to produce this thrust.

3 The first law of thermodynamics

3.1   AIMS

- To introduce the first law of thermodynamics for a closed system undergoing a cycle.
- To introduce the first law of thermodynamics for a closed system undergoing a process.
- To define heat and work as forms of energy that can cross the boundary of a closed system.
- To define the following properties of a working fluid:

  specific internal energy
  specific enthalpy
  specific entropy

- To introduce the following types of process for heat transfer to a closed system:

  isothermal process
  adiabatic process

3.2   THE FIRST LAW APPLIED TO A CLOSED SYSTEM

The first law of thermodynamics is a statement of the principle of the conservation of energy. In this chapter, only the first law for a closed system will be considered; the first law for an open system is discussed in Chapter 5.

A closed system has been defined in section 2.4.1 as a fixed quantity of fluid contained within a boundary. The boundary serves to separate the closed system from the surroundings. Only energy in the form of heat and work can cross the boundary.

3.3   FIRST LAW FOR A CYCLE

If a closed system contains a fluid and undergoes a series of processes, such that the fluid returns to its original state, the system has gone through a cycle. During the individual processes that make up the cycle, there may be transfer of heat or work. Alternatively there can be simultaneous transfer of both heat and work during a process.

Since the fluid starts at a particular state and ends the cycle at exactly the same state, there can be no net change of energy within the fluid. Therefore, for a complete cycle the net gain, or loss, of heat and work during the individual processes must cancel each other out.

This can be illustrated by the closed system shown in Figure 3.1. The system is contained within a cylinder and frictionless piston. At state 1 the system contains 1 kg of water at room temperature so that initially there is no heat transfer between the system and the surroundings. The piston is loaded with a weight so that the pressure inside the system is greater than the pressure of the surroundings.

From state 1, heat is transferred to the system causing the water to evaporate and expand. Expansion causes the piston, and weight, to be raised. Since the pressure inside the system is greater than the atmospheric pressure of the surroundings, there is a resultant force on the piston. At state 2 the piston has moved a distance x and work has been done on the surroundings. The work is given by

$$w = \text{force on piston} \times x$$

In order to achieve this work a quantity of heat q has to be transferred to the system. (Note that the quantities q and w are not equal for this process.) When the fluid reaches state 2, the heat input is stopped and the system is allowed to cool back down to room temperature.

The process between states 1 and 2 is reversible so that, on cooling, the piston moves back to its initial position. To achieve this, the surroundings must do work w on the system. At the same time there is a transfer of heat q from the system to the surroundings. Having reached state 1 again the system will have completed a cycle. During the cycle the net change in heat and work across the boundary are the same.

This is true for all closed systems irrespective of whether the net changes in heat and work are zero or not, and can be summarized in the first law of thermodynamics for a cycle:

'When a closed system undergoes a cycle then the net heat transfer to the system is equal to the net work done on the surroundings'.

Using symbols, this law can be expressed as

$$q_{net} = w_{net} \tag{3.1}$$

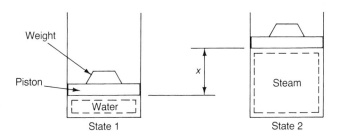

Figure 3.1 A closed system undergoing a cycle.

Example 3.1

A steam plant consists of a boiler, turbine, condenser and feed pump. For each kg of fluid circulating around the plant, 2100 kJ of heat is transferred to the boiler, the turbine does 500 kJ of work and the work input to the pump is 10 kJ. Calculate the heat transferred from the condenser.

Conceptual model

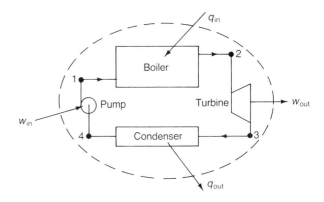

Analysis – applying equation (3.1) for the cycle

$$q_{12} + q_{34} = w_{23} + w_{41}$$
$$2100 + q_{out} = 500 - 10$$
$$q_{out} = -1610 \text{ kJ/kg}$$

3.4 FIRST LAW FOR A PROCESS

The previous section considered the application of the first law of thermodynamics to a cycle. Implicit in the discussion were the assumptions:

heat input to the system is positive;
heat output from the system is negative;
work output from the system is positive;
work input to the system is negative.

Using this notation it is now possible to consider the first law of thermodynamics as applied to a process. From the conservation of energy principle the first law of thermodynamics for a closed system undergoing a process can be expressed as

Energy entering – Energy leaving = Change of energy in
    system   system   system

The only forms of energy that can enter or leave the system are heat and work. For a unit mass of fluid within the system, heat and work are designated by q and w respectively. The first law of thermodynamics for a process can, therefore, be expressed:

$$q - w = \Delta e \tag{3.2}$$

where Δe is the net change of specific energy per unit mass of fluid, within the system.

The change of energy within the system can be assumed to consist of three parts:

kinetic energy, ΔKE;
potential energy, ΔPE;
internal energy, Δu.

Of these, the most significant is the change of internal energy Δu as the changes in kinetic energy and potential energy within a closed system can be assumed to be negligible.

Two typical examples of closed systems in which a process can take place are:

a rigid sealed container;
a cylinder and piston assembly.

For the former there is no movement of the boundary and so, by definition, there can be no change of kinetic energy or potential energy for a process in which the fluid inside the system moves from one equilibrium state to another.

Even when there is movement of the boundary, as in the case of a cylinder and piston assembly, the change in height and the velocity of the piston lead to only small changes in the kinetic and potential energy. Therefore, equation (3.1) can be expressed in the form

$$q - w = \Delta KE + \Delta PE + \Delta u$$

where

$$\Delta KE \approx 0$$
$$\Delta PE \approx 0$$

and

$$q - w = \Delta u \tag{3.3}$$

Equation (3.3) is true irrespective of whether the process is reversible or not, and is a general expression for the first law for a process.

Alternatively, the change in internal energy can be written:

$$\Delta u = u_2 - u_1$$

and incorporating this in equation (3.3) gives

$$q - w = u_2 - u_1 \tag{3.4}$$

where u_1 and u_2 are values of specific internal energy at the start and end of a process.

The internal energy of a fluid depends on its temperature (and, in the case of a vapour, on its pressure), and is a property of the fluid. This is discussed in greater detail by Look and Sauer (1988). The value of the internal energy depends on the datum used. Since it is the **change** of internal energy that is important the choice of datum can be quite arbitrary.

Example 3.2

A sealed rigid vessel contains 1 kg of fluid that is being simultaneously heated and stirred by a rotor inside the vessel. Heat transfer to the fluid is 400 kJ/kg while the rotor does 50 kJ/kg of work on the fluid. If the specific internal energy of the fluid is 400 kJ/kg at the start of the process, determine the final value.

Conceptual model

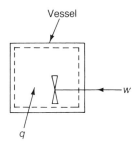

Note that since the vessel is sealed it acts as a closed system.

Analysis – assume:

1. zero change of potential energy;
2. the fluid starts and ends in a state of equilibrium with zero change of kinetic energy.

Using the values

$$q = + 400 \text{ kJ/kg}$$
$$w = - 50 \text{ kJ/kg}$$
$$u_1 = 400 \text{ kJ/kg}$$

and substituting in equation (3.4),

$$400 - (-50) = u_2 - 400$$
$$u_2 = 850 \text{ kJ/kg}$$

3.5   WORK

Work is a mechanical form of energy that can cross the boundary of a closed system. It is associated with a force moving through a distance:

$$w = F \times x$$

where F is the force on a body displaced by a distance x.

The form of work depends on whether the boundary of a closed system is static or movable.

3.5.1   Work across a static boundary

A typical example of a closed system with a static boundary is fluid inside a rigid sealed container. Work can cross the boundary by means of a rotating shaft that drives a propeller or paddle wheel inside the container, as shown in Figure 3.2.

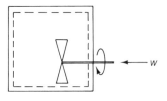

Figure 3.2 Shaft work to a closed system.

The paddle wheel imparts a velocity and therefore a kinetic energy to the fluid. This energy is dissipated to the fluid in the system through friction. Once the rotation of the paddle wheel has stopped the internal movement in the fluid will stop and the system achieves a state of equilibrium. The work input will show in the form of an increase in internal energy and equation (3.4) will still apply.

If the torque on the paddle wheel is $F \times r$, where F is a force acting at a radius r, then the work can be expressed by

$$w = F \times \text{distance}$$
$$= F \times (2\pi r N)$$

where N is the number of revolutions of the wheel. This expression can be stated in the form:

$$w = \text{torque} \times 2\pi N \tag{3.5}$$

The work done by a paddle wheel on a fluid inside a closed system is not reversible. By doing work on a fluid the internal energy is increased. A reversal of the process in which internal energy is reduced does not necessarily cause motion of the fluid to drive the paddle wheel, but could be achieved by heat transfer across the boundary to the surroundings.

Example 3.3

A sealed rigid vessel contains 1 kg of fluid with a specific internal energy of 250 kJ/kg. Work is done by means of a paddle wheel rotating with a torque of 0.1 Nm and completes 800 revolutions. Find the final value of specific internal energy if the heat loss to the surroundings is negligible.

Conceptual model

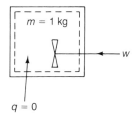

Analysis – assume:

1. $q = 0$
2. $\Delta KE = 0$

Re-arranging equation (3.4) and incorporating assumption 1:

$$u_2 = u_1 - w$$

Calculating work, from equation (3.5),

$$w = -(0.1 \times 2\pi \times 800) = -502.7 \text{ kJ/kg}$$

Since $u_1 = 250$ kJ/kg,

$$u_2 = 250 - (-502.7) = 752.7 \text{ kJ/kg}$$

Note – the negative sign for the work indicates work into the system.

46 | The first law of thermodynamics

3.5.2 Work across a moving boundary

A typical example of a closed system with a movable boundary is a cylinder and piston assembly. As the piston moves the work done is due to the force on the piston, as a result of the pressure of the fluid within the system, moving through a distance.

Consider the situation shown in Figure 3.3. One kg of fluid (say a gas as this is probably the easiest to visualize), is contained within a cylinder having a frictionless piston. The piston is loaded with several weights, numbered 1, 2, etc. With all the weights in position, as shown in Figure 3.3(a), the pressure inside the system is P. Under these conditions it can be assumed that the system is in equilibrium and there is no heat transfer between the system and the surroundings.

The removal of weight 1 will cause a sudden drop in pressure to P_1 and the piston to move. Once the system has regained a state of equilibrium the piston will have move through a distance δx_1, as shown in Figure 3.3(b). Similarly, removal of weight 2 will cause a drop in pressure to P_2 and a movement of the piston by δx_2, as shown in Figure 3.3(c). Movement of the piston causes work to be done and, ignoring the pressure of the surroundings, the work done by removing weight 1 is given by:

$$w = P_1 \times A \times \delta x_1 = P_1 \times (A \delta x_1)$$

where A is the area of the piston.

Now $(A \delta x_1)$ is a change in volume and since the system contains unit mass, this is a change in specific volume, δv_1. Hence, the work done by removing weight 1 is given by

$$w = P_1 \times \delta v_1$$

Similarly, for weight 2, the work is

$$w_2 = P_2 \times \delta v_2$$

The expansion processes for the removal of the weights is shown on a pressure–specific volume diagram in Figure 3.4. If, for each individual process, the system is at equilibrium then the total work for all processes is the sum of all the individual works:

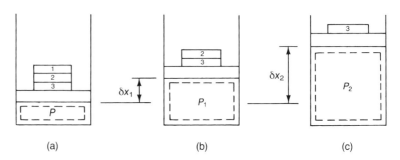

Figure 3.3 A piston loaded with weights.

$$w = w_1 + w_2 + \dots$$
$$= P_1 \delta v_1 + P_2 \delta v_2 + \dots$$

Also, the magnitude of the work is equal to the area, $P \delta v$, on the diagram for each process.

If each incremental weight and, therefore, the resultant change in volume was made infinitesimally small, the work for each incremental change in pressure would be

$$dw = Pdv$$

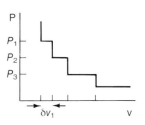

Figure 3.4
Pressure–specific volume diagram of the expansion process.

With such small changes each individual process becomes merged into one continuous process. For the whole process, as shown in Figure 3.5, the total work is the integration of all the infinitesimal changes between the initial state, point 1, and the final state, point 2:

$$w = \int_1^2 Pdv \qquad (3.6)$$

Since the original series of processes, portrayed in Figure 3.3, are reversible it is assumed that the continuous process, shown in Figure 3.5, is also reversible. In fact, equation (3.6) is only valid for a reversible process. If there was friction between the piston and cylinder then some part of the work would be used in overcoming that friction and the actual work done would be less.

Just as for the original series of processes, the work is the area under the curve from point 1 to point 2. It can, therefore, be concluded that for a reversible process, within a closed system, the area under a process curve on a pressure–specific volume diagram is equal to the work done. This is true irrespective of whether there is heat transfer across the boundary or not.

The principle holds true if the mass of fluid inside the closed system is greater, or less, than 1 kg. By multiplying both sides of equation (3.6) by the mass of fluid m it can be shown that the actual work done is equal to the area under the process curve on a pressure–volume (P–v) diagram. Under these circumstances the actual work done is expressed in joules (or kJ).

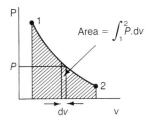

Figure 3.5 Continuous expansion process.

Example 3.4
A fluid at a pressure of 500 kPa and specific volume 0.1 m³/kg, is contained within a cylinder by a piston. If the piston moves until the specific volume is 0.5 m³/kg and the pressure remains constant throughout the process, calculate the work done per unit mass of fluid.

Conceptual model – It would be possible to draw a cylinder and piston but a more useful model is the process diagram.

Process diagram

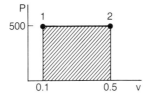

Analysis – assume:

1. the movement of the piston is frictionless;
2. the process is reversible.

It is possible to apply equation (3.6):

$$w = \int_1^2 P.\,dv$$

= area under curve
= $P(v_2 - v_1)$
= 500 (0.5 – 0.2) = 200 kJ/kg
      0.1

Since the work is being done on the surroundings, the value of work is positive.

Example 3.5
A fluid at a pressure of 200 kPa and specific volume 0.6 m³/kg is compressed within a cylinder by a piston until the specific volume is 0.2 m³/kg. If the fluid obeys the law Pv^2 = constant, calculate the work done.

Process diagram

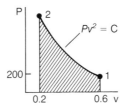

Analysis – as for example 3.4, assume that equation (3.6) is valid:

$$w = \int_1^2 P.\,dv$$

now

$$P = \frac{\text{constant}}{v^2}$$

and, from the initial conditions,

$$\text{constant} = 200 \times (0.6)^2 = 72 \text{ kJ/kg}$$

Substituting in equation (3.5)

$$w = 72 \int_{v_1}^{v_2} \frac{dv}{v^2}$$

$$= 72 \left[-\frac{1}{v} \right]_{0.6}^{0.2}$$

$$= 72 \left[\left(-\frac{1}{0.2} \right) - \left(-\frac{1}{0.6} \right) \right] = -240 \text{ kJ/kg}$$

Note – The negative sign indicates work being done on the system.

3.5.3 Work done during a cycle

When a closed system undergoes a series of processes so that the fluid inside the system returns to its original state, then the system has gone through a cycle. If each of the individual processes making up the cycle are reversible, then the whole cycle is reversible. Following the discussion in the preceding section, if each of the processes making up the cycle are plotted on a P–v diagram, then the work done during each process will be the area under each process curve.

Figure 3.6 shows a cycle made up of four reversible processes 1 to 2, 2 to 3, 3 to 4 and 4 back to 1. Because each of the processes is reversible the whole cycle is reversible. During the first two processes, 1–2–3, work is being done on the surroundings and the total quantity is the area under the curve 1–2–3.

During the last two processes, 3–4–1, work is being done on the system and the quantity of this work is the area under curve 3–4–1. Summing the areas shown in Figure 3.6 gives

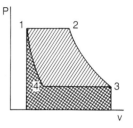

Figure 3.6 P–v diagram of a cycle.

Area enclosed = Area under – Area under
by 1–2–3–4–1   1–2–3   3–4–1

It follows that the enclosed area within the cycle 1–2–3–4–1 represents the net work for the cycle. Since the area under curve 1–2–3 is greater than that under curve 3–4–1, the net work is an output from the system and using the sign convention, the net work from this cycle is positive.

As the cycle is reversible it is possible to reverse the direction of each process so that the cycle works in the order 1–4–3–2–1. Under these circumstances the work done is on the fluid in the system and the net work of the cycle is negative.

It follows that a cycle following a clockwise direction on a P–v diagram produces a positive work output, whilst a reversed cycle following an anti-clockwise direction represents a negative work input.

Example 3.6

A fluid inside a closed cycle undergoes a series of reversible processes. The fluid initially starts at 800 kPa and specific volume of 0.05 m³/kg. It is allowed to expand to a specific volume of 0.25 m³/kg at constant pressure. At this volume the pressure is then reduced to 200 kPa. At this pressure the fluid is compressed until the specific volume is again 0.05 m³/kg. Finally, the pressure is increased to 800 kPa and the fluid reaches its initial state. Find the specific work done during this cycle.

Conceptual model – the question does not define the type of closed system used. A P–v diagram of the processes is sufficient to describe the whole cycle.

Process diagram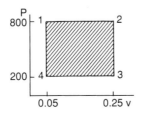

Analysis – since each process is reversible, the net work for the cycle is the area contained by the processes 1–2–3–4–1.

Process 1–2 $\quad w_{12} = P_1 (v_2 - v_1)$
$\qquad\qquad\quad = 800 \, (0.25 - 0.05) = 160 \, \text{kJ/kg}$
Process 2–3 $\quad w_{23} = 0$ (because the change in volume is zero)
Process 3–4 $\quad w_{34} = P_3 (v_4 - v_3)$
$\qquad\qquad\quad = 200 \, (0.05 - 0.25) = -40 \, \text{kJ/kg}$
Process 4–1 $\quad w_{41} = 0$

Summing the work done for each process,

$$w_{net} = w_{12} + w_{23} + w_{34} + w_{41}$$
$$= 160 + 0 - 40 + 0 = 120 \, \text{kJ/kg}$$

Note – the net work is positive, signifying a work **output** from the system.

3.6   HEAT

Heat is a form of energy that can cross the boundary of a system. (Note that the type of system is not specified as heat can be transferred to both closed and open systems.) As such, it is analogous to work but, whereas work is easier to visualize because it is associated with physical movement of a piston or a rotating shaft, heat is more difficult to understand.

This understanding is not helped by the inherent confusion within the English language about heat and temperature. When something is said to be hot what is implied is that the heat content of the object is high. But this is impossible because an object (system) cannot contain heat, as heat is not a property. What is really meant is that the temperature of the object is high. Alternatively, having now been introduced to the first law of thermodynamics, it would be possible to relate the 'hotness' of an object, or system, to its internal energy.

Nevertheless, even though heat cannot be seen in a physical sense, the effect of the transfer of heat to or from a system is experienced on an everyday basis. A cup of tea, if left to stand long enough, will cool down to room temperature. An ice cube, taken from a refrigerator, will eventually melt and the temperature of the resulting water rise to that of the room. In both these examples the direction of the transfer of heat is from the higher temperature body to the lower temperature one.

Heat can, therefore, be defined as 'the form of energy that is transferred between a system and its surroundings as a result of a temperature difference'. There can only be a transfer of energy across the boundary in the form of heat if there is a temperature difference between the system and its surroundings. Conversely, if the system and surroundings are at the same temperature there is no heat transfer across the boundary.

Strictly speaking, the term 'heat' is a name given to the particular form of energy crossing the boundary. However, heat is more usually referred to in thermodynamics through the term 'heat transfer', which is consistent with the ability of heat to raise or lower the energy within a system.

There are three modes of heat transfer:

convection
conduction
radiation

All three are different. Convection relies on movement of a fluid. Conduction relies on transfer of energy between molecules within a solid or fluid. Radiation is a form of electromagnetic energy transmission and is independent of any substance between the emitter and receiver of such energy. However, all three modes of heat transfer rely on a temperature difference for the transfer of energy to take place.

If the temperature of the surroundings is higher than the fluid in the system then the transfer of energy is positive and there is said to be 'heat addition' to the system. If the fluid in the system is at a higher temperature than the surroundings, then there is 'heat rejection' from the system.

3.6.1 Temperature-specific entropy diagram

In section 3.3 it was shown that if a closed system undergoes a cycle then the net heat transfer to the system is equal to the net work done on the surroundings. Also, in section 3.5.3 it was shown that the net work for a reversible cycle is the area of that cycle when plotted on a P–v diagram. It can be stated that

$$q_{net} = w_{net} = \int_c P.dv$$

where the subscript 'c' under the integral sign means for the whole cycle.

This allows the magnitude of the net heat transferred to be found but says nothing about the heat transfer for each individual process. This is because there is a difference between the heat transferred and the work done for each process, that difference being the change of specific internal energy during each process.

It is necessary to create another process diagram, different to the P–v diagram, to describe heat transfer during a process. From the preceding discussion, heat transfer depends on a temperature difference and therefore the temperature T should be one of the properties used for any such process diagram. The question that remains is which other property can be used?

Neither specific volume v or specific internal energy u is valid as the former is used to evaluate the work done, whilst the latter is a function of both the heat transfer and the work done.

Clearly what is required is a property that is only a function of the heat transferred q and the temperature of the fluid within the system T. That property is called entropy or, when applied to 1 kg of fluid, specific entropy and is given the symbol s.

The word entropy derives from the Greek, meaning transformation. In other words, an alteration to the state of a fluid inside a system. A fluid at high temperature, having molecules moving about in a highly disordered manner, has a higher value of specific entropy than one in which the molecules move in a more ordered manner. Entropy is, therefore, sometimes referred to as a measure of the disorder within a fluid or within a system.

It is sufficient at this stage to simply consider specific entropy as a property. Just as work is the area under a reversible process curve on a P–v diagram, heat is the area under a reversible process curve on a T–s diagram and the quantity of heat transferred is given by

$$q = \int_1^2 T \, ds \tag{3.7}$$

Checking on the units, q is in kJ/kg and T is in K. Therefore, the units for specific entropy s are in kJ/kg K.

These are exactly the same units as for specific heat, but the two properties are **totally** different and should not be confused. Specific heat is a measure of a fluid's capacity for storing energy and is defined as 'the energy required to raise the temperature of a unit mass of a substance by one degree'. Specific entropy, on the other hand, is related to the heat that can be transferred across the boundary to a unit mass of a substance when it is at a particular temperature. It describes the **quality** of heat, because a **quantity** of heat transferred at a high temperature will have a smaller value of entropy than the same **quantity** transferred at a lower temperature.

3.6.2 Isothermal process

An isothermal process is one in which the heat transfer to or from a system takes place at constant temperature. Consider the closed system shown in Figure 3.7, contained within a cylinder having a frictionless piston. The pressure inside the closed system is constant, being determined by the weight on top of the piston. If the closed system contains 1 kg of water at the boiling point for the given pressure, then any heat addition to the system will cause the water to evaporate. As a result the vapour will cause the piston to move upwards, doing work on the surroundings. Since the pressure is constant, the temperature at which the evaporation takes place will be constant and the process is isothermal.

The process is shown on the T–s diagram in Figure 3.7 as a horizontal line between points 1 and 2. If the heat input is removed and the system is allowed to

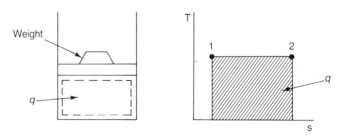

Figure 3.7 System undergoing an isothermal process.

reject heat to the surroundings, the vapour will condense and the process will be reversed from the state at point 2 back to the state at point 1. As a consequence, the heat rejected to the surroundings will also equal the heat addition q.

Since the isothermal process 1 to 2 is reversible, it follows that the area under the line 1–2 on the T–s diagram, is equal to the quantity of heat transferred, q.

Example 3.7

A fluid at a temperature of 400 K and specific entropy 2 kJ/kg K, is contained within a cylinder by a frictionless piston. Heat is added by means of an isothermal process until the specific entropy reaches a value of 6 kJ/kg K. Calculate the quantity of heat transferred to the system during this process.

Conceptual model – a process diagram is sufficient to model the situation.

Process diagram

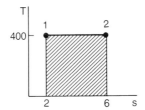

Analysis – assume the isothermal process to be reversible. Then, using equation (3.7)

$$q = \int_{1}^{2} T.ds$$

$$= \text{area under curve}$$
$$= T (s_2 - s_1)$$
$$= 400 (6 - 2) = 1600 \text{ kJ/kg}$$

Since the heat is being added to the system the value of the heat transferred is positive.

3.6.3 Adiabatic process

An adiabatic process is one in which there is no heat transfer to or from a system, during a process. This can be achieved in one of two ways.

1. By maintaining the temperature of the fluid inside the system at the same temperature as the surroundings.
2. By insulating the system so that the heat transfer across the boundary is negligible.

The two ways of achieving an adiabatic process are considered in greater detail in Figure 3.8. The closed system in Figure 3.8(a) is contained within a cylinder and frictionless piston. If the fluid inside the closed system is a liquid that is evaporating under constant pressure, the temperature inside the system, T_i, will be constant. If T_i is equal to the temperature of the surroundings T_o,

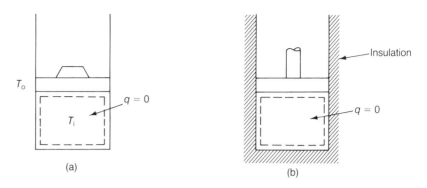

Figure 3.8 Achieving an adiabatic process.

there is no temperature difference across the boundary and so heat transfer cannot take place. Such a situation is entirely feasible if the fluid inside the closed system is, say, a boiling refrigerant, since such fluids can boil at room temperature provided the pressure inside the system is greater than atmospheric.

The closed system shown in Figure 3.8(b) is representative of a general adiabatic situation because, irrespective of the fluid inside the system, the insulation can be assumed to prevent the transfer of heat between the surroundings and the system.

Figure 3.9 shows a general reversible adiabatic process as plotted on both the P–v and T–s diagrams. Although the heat transferred is zero there must be work done during a process. This can be shown from the first law of thermodynamics for a process, equation (3.4):

$$q - w = u_2 - u_1$$

Since $q = 0$, the process can only be accomplished by work crossing the boundary. The quantity of the work is equal to the change in the specific internal energy:

$$w = u_1 - u_2$$

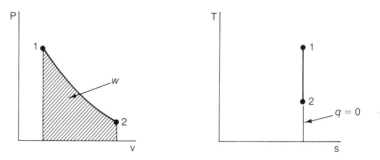

Figure 3.9 A reversible adiabatic process.

The work is shown as the area under the curve on the P–v diagram. The equivalent process on the T–s diagram is a vertical straight line, from state 1 to state 2. Since the heat transferred is zero it follows that the area under the line from 1 to 2 on the T–s diagram is zero.

Because there is no change of entropy, a reversible adiabatic process is also referred to as an 'isentropic' process, i.e. at constant entropy.

It must be stressed that the T–s diagram in Figure 3.9 is only valid for a **reversible** adiabatic process. It is possible to have adiabatic processes that are irreversible, but the fact that they are irreversible means that equation (3.7):

$$q = \int_1^2 T \, . \, ds$$

is no longer valid, and the area under any resulting curve on the T–s diagram **no longer** relates to the heat transferred.

An irreversible adiabatic process is shown in Figure 3.10. The closed system in Figure 3.10(a) consists of two containers joined by a pipe and valve. Since the system is insulated any process must be adiabatic. Container A holds a fluid that can expand, i.e. either a vapour or a gas. Container B has a vacuum. When the valve is opened the fluid flows through the pipe to fill both containers. In the process the volume of fluid is increased and the pressure, and temperature, are decreased. The process is irreversible because there is no mechanism to move all the fluid back into Container A.

Figure 3.10(b) shows the process on the T–s diagram. The resultant curve is no longer a vertical straight line but indicates an increase in entropy. This is true of any irreversible process. As a general rule, there is a tendency for entropy to increase during an irreversible process.

The heat transfer during the adiabatic process is zero, so the area under the curve from 1 to 2 in Figure 3.10(b) no longer has any meaning. Similarly, if the process was plotted on a P–v diagram the resulting area under the curve would not relate to work, as the closed system in Figure 3.10(a) has no means of doing work on the surroundings.

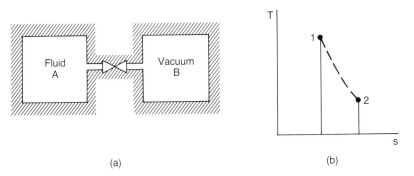

(a)   (b)

Figure 3.10 An irreversible adiabatic process.

3.6.4 Heat transferred during a cycle

When a closed system undergoes a reversible cycle, the individual processes making up the cycle can be plotted on the T–s diagram. The resulting area enclosed within the cycle is the quantity of heat transferred to, or from, the system during the cycle.

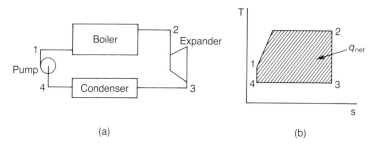

(a) (b)

Figure 3.11 Cycle of a steam plant.

Figure 3.11(a) shows the steam plant originally introduced in Figure 2.2. Assuming that the plant operates on a reversible cycle, the individual processes can be plotted on the T–s diagram, as shown in Figure 3.11(b), as a way of quantifying the net heat transferred.

Between 1 and 2 the temperature of the water leaving the pump is increased until it reaches boiling point, thereafter the evaporation continues in the boiler at constant temperature. From 2 to 3 it can be assumed that the steam expands in an adiabatic process. This is a realistic assumption because the expander, either a reciprocating engine or rotating turbine, can be insulated to prevent any significant heat being lost to the surroundings. From 3 to 4 the steam exhausting from the expansion process is condensed at constant pressure and, therefore, at constant temperature. Finally, the water leaving the condenser at 4 is pumped back into the boiler by means of an adiabatic process to 1. The cycle, shown on the T–s diagram in Figure 3.11(b), can be compared with the equivalent cycle, shown on the P–v diagram in Figure 2.12(b).

The net heat transferred during the cycle is given by the area enclosed within the state points 1–2–3–4–1 and is heat transferred to the system. Using the sign convention, the net heat for this cycle is positive.

It follows that a cycle following a clockwise direction on a T–s diagram has a positive heat input, while a reversed cycle following an anti-clockwise direction represents a negative heat output.

From the first law for a cycle, the net heat transfer to a closed system is equal to the net work done on the surroundings. This can be expressed as

$$\int_c T.\mathrm{d}s = \int_c P.\mathrm{d}v \tag{3.8}$$

provided that the cycles are made up of reversible processes.

Example 3.8
A closed system works on a cycle in which a fluid is initially heated from a temperature of 300 K and specific entropy of 1.5 kJ/kg K, through a linear

process to a temperature of 500 K and specific entropy of 2.5 kJ/kg K. The fluid is then expanded adiabatically to the original temperature of 300 K. Finally, the fluid undergoes an isothermal process back to a specific entropy of 1.5 kJ/kg K.

Assuming that the cycle is reversible, calculate the work done during this cycle, per unit mass of fluid.

Conceptual model – the question does not define the type of closed system used. A T–s diagram of the processes is sufficient to describe the whole cycle.

Process diagram

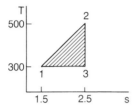

Analysis – apply equation (3.1):

$$w_{net} = q_{net}$$

Since each process is reversible, the net heat for the cycle is the area contained by the processes 1–2–3–1.

Process 1–2 $q_{12} = \left(\dfrac{T_1 + T_2}{2}\right)(s_1 - s_2)$

$$= \left(\frac{300 + 500}{2}\right)(2.5 - 1.5) = 400 \text{ kJ/kg}$$

Process 2–3 $q_{23} = 0$ (by definition)
Process 3–1 $q_{31} = T_1 (s_1 - s_2)$
$$= 300 \,(1.5 - 2.5) = -300 \text{ kJ/kg}$$

Summing the heat transferred for each process,
$$q_{net} = q_{12} + q_{23} + q_{31}$$
$$= 400 + 0 - 300 = 100 \text{ kJ/kg}$$
and $w_{net} = q_{net} = 100 \text{ kJ/kg}$

3.7   CONSTANT PRESSURE PROCESS

It is appropriate at this stage to apply the first law of thermodynamics to a process to introduce one further thermodynamic property, namely 'enthalpy'. For this it is convenient to consider a constant pressure process within a closed system.

Consider the closed system shown in Figure 3.12(a) in which a fluid is contained within a cylinder having a frictionless piston. Since the piston is loaded with a weight, the process that the fluid undergoes will be at constant pressure, as shown on the P–v diagram in Figure 3.12(b).

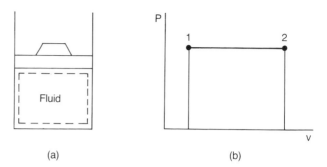

Figure 3.12 A constant pressure process.

Applying the first law for a process, from equation (3.4),

$$q - w = u_2 - u_1$$

Assuming that the process from 1 to 2 is reversible, then the work done is the area under the curve on the P–v diagram:

$$w = P \, (v_2 - v_1)$$

substituting this work into the previous equation gives

$$
\begin{aligned}
q &= (u_2 - u_1) + w \\
&= (u_2 - u_1) + P \, (v_2 - v_1)
\end{aligned}
$$

Re-arranging gives

$$q = (u_2 + Pv_2) - (u_1 + Pv_1)$$

Since u, P and v are all properties, it follows that $(u + Pv)$ is also a property. This property is the specific enthalpy, given the symbol h.

Therefore, for a reversible constant pressure process in a closed system,

$$q = h_2 - h_1 \tag{3.9}$$

where $h = u + Pv$.

The units of both u and Pv are consistent and specific enthalpy is expressed in kJ/kg.

Enthalpy is more generally encountered in flow situations, discussed in Chapter 5. However, it is introduced here to provide a basis for the properties of fluids discussed in the next chapter. Just as the change of internal energy is important for a process in a closed system, so the change of enthalpy is important in an open system. A datum, at which specific enthalpy has a value of zero, can be chosen in a quite arbitrary manner.

3.8   RATE OF HEAT TRANSFER

The first law deals with the quantity of heat that is transferred to, or from, a closed system. But this in no way says anything about the rate at which the heat is transferred.

From the discussion in section 3.6 it was explained that heat can only be transferred as a result of a temperature difference. The greater the temperature difference the more rapid will the heat be transferred. Conversely, the lower the temperature difference, the slower the rate at which heat will be transferred.

This can be illustrated by the situation shown in Figure 3.13. A closed system (a) is shown in which the fluid inside the system is at temperature T_i and the surroundings outside are at temperature T_o. If T_o is greater than T_i there is a temperature difference $(T_o - T_i)$ that will cause heat to be transferred to the system.

If a quantity of heat q is transferred to the system such that the fluid inside increases from temperature T_1 to T_2 the rate at which the temperature increases is illustrated by Figure 3.13(b). This shows a plot of temperature of the fluid against time. Initially, the temperature difference $(T_o - T_i)$ causes rapid heat transfer to the system and the temperature of the fluid increases rapidly. As the temperature of the fluid approaches the outside temperature, the temperature difference $(T_o - T_i)$ decreases and the rate of heat transfer is reduced.

Heat transfer is therefore a time-dependent process. The rate of heat transfer can be defined as the quantity of heat transferred during a given time. However, this is only true for a process if the quantity of heat transferred and the time are taken as small increments. For a system containing 1 kg of fluid, the rate of heat transfer Q is given by

$$Q = \frac{\delta q}{\delta t}$$

Alternatively, for a system containing a mass m of fluid,

$$Q = m\frac{\delta q}{\delta t} \tag{3.10}$$

Introducing the units for the variables:

$$kg\frac{j/kg}{s} = \frac{j}{s} = W$$

Therefore, the units for the rate of heat transfer are watts.

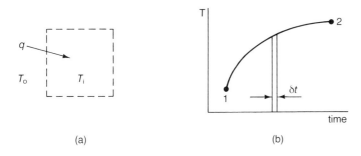

(a) (b)

Figure 3.13 Rate of heat transfer to a system.

This brief discussion of the rate of heat transfer is presented here as it follows on from the discussion of heat, section 3.6. Also, it is important to have a clear idea of the difference between the rate of heat transfer Q, and the quantity of heat transferred q. The rate of heat transfer is used in the analysis of heat transfer situations, presented in Chapter 12.

SUMMARY

In this chapter the application of the first law of thermodynamics to a closed system has been discussed. The key terms that have been introduced are:

first law of thermodynamics
work
heat
net heat for a cycle
net work for a cycle
specific internal energy
specific enthalpy
specific entropy
isothermal process
adiabatic process
rate of heat transfer

Key equations that have been introduced.

For a cycle:

$$q_{net} = w_{net} \tag{3.1}$$

For a process:

$$q - w = u_2 - u_1 \tag{3.4}$$

For a reversible process:

$$w = \int_1^2 P \mathrm{d}v \tag{3.6}$$

$$q = \int_1^2 T \mathrm{d}s \tag{3.7}$$

Defining enthalpy:

$$h = u + Pv \tag{3.9}$$

PROBLEMS

1. Sunlight enters a room through a window. Considering the room as an insulated closed system, will there be any change in the internal energy of the room?

2. In a kitchen, a refrigerator is connected to the electric mains and has its door wide open. Considering the kitchen as an insulated closed system, will the internal energy rise, fall or remain the same?

3. A closed system undergoes a cycle consisting of two processes. During the first process 40 kJ of heat is transferred to the system and 30 kJ of work is done on the surroundings. During the second process 45 kJ of work is done on the system. Find the heat transferred during the second process.

4. Complete the following table for processes inside a closed system. All values are in kJ/kg.

Process	q	w	u_1	u_2	Δu
a	20	15	35		
b		−8	12		20
c	−10			6	−12
d	25		18		10

5. One kg of fluid at a pressure of 250 kPa is compressed from a volume of 0.8 m^3 to 0.2 m^3 within a cylinder and piston assembly. Calculate the work done on the fluid.

6. A balloon is inflated by filling it with a gas from a high-pressure cylinder. If the balloon is initially fully deflated and is inflated to a volume of 0.01 m^3, determine the work done during this process. Take the atmospheric pressure as 100 kPa.

7. One kg of fluid is expanded from 600 kPa down to 200 kPa according to the linear relationship

$$P = av + b$$

where a and b are constants. If $a = -1000$ kPa kg/m^3 and the initial volume of the fluid is 0.2 m^3/kg, calculate the work done during the process.

8. Consider the expansion process defined in question 7 as part of a cycle. At the end of the expansion the fluid undergoes a constant pressure process to a pressure of 200 kPa and specific volume of 0.2 m^3/kg. The fluid is then compressed at constant volume to a pressure of 600 kPa. Find the net work and hence determine the net heat addition for the cycle.

9. A closed system undergoes a cycle in which heat is added isothermally at a temperature of 500 K with a change of specific entropy of 2 kJ/kg K. If the rest of the cycle consists of an adiabatic expansion down to a temperature of 300 K, isothermal heat rejection to the initial value of entropy and, finally, adiabatic compression to the initial temperature, find the net heat transferred during the cycle.

4 Fluid properties

4.1 AIMS

- To introduce the types of fluid used in thermodynamic processes.
- To explain the changes that take place during a two-phase process.
- To define the following terms for a two-phase fluid:

  saturated conditions
  wet vapour
  critical point
  dryness fraction

- To introduce the use of tables to find the properties of vapours.
- To define the equation of state for a perfect gas.
- To introduce the specific heats at constant volume and constant pressure for a perfect gas.
- To define the specific heats for a perfect gas and the relationship between them.
- To define the basic relationships for a perfect gas undergoing the following processes:

  isothermal
  adiabatic

- To introduce the ratio of the specific heats for a perfect gas, undergoing an adiabatic process.

4.2 FLUIDS USED IN THERMODYNAMIC PROCESSES

The preceding chapter described the behaviour of closed systems undergoing a process or cycle. Without specifically stating so, it was assumed that the fluid within the system was a pure substance. This is defined as a fluid that has a fixed chemical composition throughout the thermodynamic changes.

A pure substance does not have to be a single element. A mixture of various elements or compounds can be considered as a pure substance provided that the overall composition does not change. Air is a fluid that is widely employed in thermodynamic processes or devices. As such, it is taken to be a pure substance, even though it is made up of several gases including oxygen, nitrogen and carbon dioxide. It is assumed that the composition remains constant and that none of the constituent gases undergoes a chemical change during a thermodynamic process.

The types of fluid used within thermodynamic processes are liquids, vapours or gases. It is possible for one substance to exist in all these different states. A typical example is water. Under normal atmospheric conditions it exists as a liquid. Raising the temperature to its boiling point causes the water to change to steam, a vapour. When all the water has been converted to steam, additional heat input causes the temperature to rise above the boiling point into what is called the superheat region. If the temperature of the superheated steam is sufficiently high, then the steam achieves what is effectively a gaseous state.

These various states of a fluid are called 'phases'. Fluids tend to be referred to by the phase in which they are at equilibrium under normal atmospheric conditions. Air, oxygen and hydrogen, for example, are considered as gases. Water is considered as a liquid since, at atmospheric pressure, the temperature has to be well above ambient to cause boiling.

During boiling, or condensing, a fluid such as water or a refrigerant can consist of both liquid and vapour in equilibrium. Under these conditions the fluid is referred to as being in a 'two-phase' state.

4.3 A FLUID UNDERGOING A PHASE CHANGE

Consider a cylinder and frictionless piston assembly, as shown in Figure 4.1 (a), containing 1 kg of water at room temperature. The cylinder and piston form a closed system. With the piston loaded as shown, the pressure inside the system is greater than atmospheric and will be constant throughout the process.

As heat is transferred to the water, the temperature rises until it reaches boiling point, state 1 on the curve shown in Figure 4.1(b). Thereafter, continuing heat transfer causes boiling to take place at constant temperature until, at state 2, the water is completely changed to steam. From this condition, continuing heat input to the system causes the temperature of the steam to increase in the superheat region to state 3.

The curve shown in Figure 4.1(b) is typical of any fluid going through a phase change. Oxygen is generally considered as a gas, but at its boiling point of −183°C it goes through exactly the same phase change as any other liquid. At room temperature oxygen is so far above its boiling point that it no longer behaves like a vapour, even a superheated vapour, but as a gas.

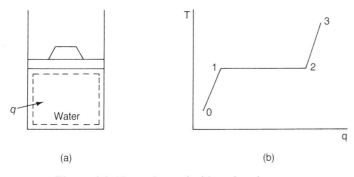

(a)   (b)

Figure 4.1 Phase change inside a closed system.

Although Figure 4.1(b) provides a useful illustration of what happens to a fluid changing from a liquid to a vapour, it is **not** a property diagram of the process. This is because the heat input q is not a property of the fluid but a measure of the quantity of energy transferred from the surroundings to the closed system. As a consequence of the first law of thermodynamics, the heat input will cause some work to be done on the surroundings but, more importantly, there will be an increase in the specific internal energy during the process.

There are changes to other fluid properties during the process. The temperature changes, as shown in Figure 4.1(b), and there will be associated changes in specific enthalpy and specific volume. However, the property that is of most interest is specific entropy, s. If the change in temperature of the water, or steam, is plotted against entropy, then the phase change can be illustrated on the T–s property diagram.

Between states 1 and 2, Figure 4.1(b), the water boils at constant temperature as a result of the heat input q_{12}. This process is reversible because the transfer of a quantity of heat, equal to q_{12} but **away** from the system, would cause the steam at 2 to condense back to the water at 1. For a reversible process, equation (3.7),

$$q_{12} = \int_1^2 T \mathrm{d}s$$

Since the temperature is constant between points 1 and 2, it follows that

$$q_{12} = T(s_2 - s_1)$$

Representing this on the T–s diagram gives a straight line between 1 and 2, similar to that shown in Figure 4.1(b). By similar arguments the increase in temperature of the water, 0 to 1, and the increase of temperature in the superheat region, 2 to 3, can also be represented as lines on the T–s diagram. The changes shown in Figure 4.1(b) are shown on the T–s process diagram in Figure 4.2.

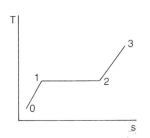

Figure 4.2 Phase change represented on a T–s diagram.

4.3.1 Two-phase fluid conditions

At the beginning of the 19th century, one of the theories regarding heat was that it was a fluid that pervaded an object or substance. When heat entered, the object or substance became 'saturated' with this colourless, odourless, weightless fluid. Even though heat has since been recognized as being a form of energy, the expression 'saturated' still remains and is used to describe the condition of a fluid undergoing a two-phase process.

At a particular pressure, a liquid will change to a vapour at one constant temperature. Although this temperature has been referred to as the boiling point, in engineering thermodynamics it is called the 'saturation temperature'. The saturation temperature is defined as the temperature at which a change of phase can take place. Clearly, for a given fluid, the saturation temperature occurs at a particular pressure. For example, water at standard atmospheric pressure has a saturation temperature of 100°C. Either the pressure or the saturation temperature can be used, as a property, to define a fluid undergoing a change of phase.

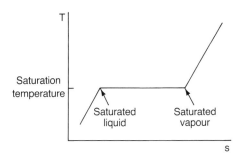

Figure 4.3 Saturation conditions.

When a liquid reaches the saturation temperature, it has reached the maximum temperature at which it can exist as a liquid. At this point it is called a 'saturated liquid'. Similarly, if all the saturated liquid is converted to vapour, then the vapour at saturation temperature is called the 'saturated vapour', Figure 4.3.

Between the saturated liquid and saturated vapour states, the fluid will be at saturated temperature but consist of a proportion of liquid and a proportion of vapour in equilibrium. This condition is termed a 'wet vapour'. The steam flowing out of the spout of a kettle is a wet vapour, otherwise it could not be seen. It is the droplets of saturated water in the steam that makes it visible.

In order for a saturated liquid to be changed to a saturated vapour, energy must enter the system in the form of heat. The amount of energy required depends on the saturation temperature of the fluid and this, in turn, depends on the pressure of the fluid. As the pressure increases, so does the saturation temperature. It is possible to draw a series of two-phase changes on the T–s diagram, as shown in Figure 4.4.

As the pressure increases the energy input necessary to change from a saturated liquid to a saturated vapour is reduced. Therefore, the entropy values of the saturated liquid and the saturated vapour draw closer together, with increasing pressure, until they are coincident. This is called the 'critical point' and represents the pressure at which saturated liquid can be instantaneously changed to saturated vapour. The critical point for water occurs at a pressure of 22.12 MPa, which is equivalent to a saturation temperature of 374.15°C.

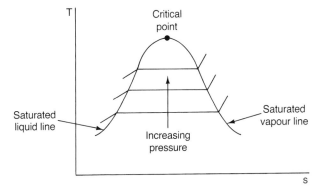

Figure 4.4 Wet vapour region.

The critical point is shown in Figure 4.4. Below it is a loop that defines the wet vapour region. On the left hand side of the loop is the saturated liquid line, the line that joins all the saturated liquid states. On the right hand side of the loop is the saturated vapour line, the line that joins all the saturated vapour states.

The state of a fluid can be defined by two independent properties. In the case of a wet vapour, both the saturation temperature and the pressure are dependent on each other, so it is only necessary to use one of these as an **independent property**. One further independent property is required to define the state of a wet vapour. This is termed the 'dryness fraction' and is given the symbol x. (Some textbooks also refer to the dryness fraction as the quality of the vapour.)

The dryness fraction is the ratio of saturated vapour in a mixture of liquid and vapour, based on mass:

$$x = \frac{\text{mass of saturated vapour}}{\text{mass of wet vapour}}$$

It follows that for every kg of wet vapour, the mass of saturated vapour is x kg and the mass of saturated liquid is $(1-x)$ kg. Hence, it follows that saturated vapour has a dryness fraction of 1 and saturated liquid has a dryness fraction of 0.

Example 4.1

One kg of wet vapour has a dryness fraction of 0.7. If the enthalpy of the saturated liquid is 417 kJ/kg and the enthalpy of the saturated vapour is 2675 kJ/kg, calculate the enthalpy of the wet vapour.

Process diagram

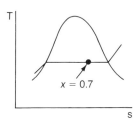

$x = 0.7$

Analysis – in 1 kg of wet vapour:

mass of saturated vapour	= 0.7 kg
mass of saturated liquid	= 0.3 kg
enthalpy of saturated vapour	= 0.7×2675 = 1872.5 kJ
enthalpy of saturated liquid	= 0.3×417 = 125.1 kJ
specific enthalpy of the wet vapour	= 1872.5 + 125.1
	= 1997.6 kJ/kg

Example 4.2

A wet vapour has a dryness fraction of 0.75. The specific volume of the saturated vapour is 0.12 m³/kg and the specific volume of the saturated liquid is 0.0017 m³/kg. Find the specific volume of the wet vapour.

Process diagram – as for example 4.1.

Analysis – in 1 kg of wet vapour:

mass of saturated vapour = 0.75 kg
mass of saturated liquid = 0.25 kg
volume of saturated vapour = $0.75 \times 0.12 = 0.09 \text{ m}^3$
volume of saturated liquid = $0.25 \times 0.0017 = 0.0004 \text{ m}^3$
specific volume of wet vapour = $0.09 + 0.0004 = 0.0904 \text{ m}^3$

Note – the specific volume of a saturated liquid is very much smaller than that of a saturated vapour. In the example above, the volume contributed by the saturated liquid was two orders of magnitude less than that for the saturated vapour. Therefore, in calculating the specific volume of a wet vapour, the volume of the saturated liquid is generally ignored – except at higher pressures, i.e. greater than 5000 kPa.

4.3.2 Superheat vapour conditions

Once a fluid has reached the saturated vapour condition, any further heat input will cause the temperature of the vapour to rise into the superheat region. Figure 4.5 shows the variation of temperature in the superheat region for a fluid under constant pressure conditions.

The temperature of a superheated vapour is independent of the pressure. Therefore, pressure and temperature can be used to define the state of a superheated vapour, as they represent two independent properties of the fluid.

4.4 USE OF VAPOUR TABLES

The properties of wet, saturated and superheated vapours can be found from tables of thermodynamic properties. Since the properties of steam are most widely used in thermodynamic analysis, these tables tend to be colloquially referred to as 'steam tables'. However, this is a misnomer as most modern tables of properties cover a wider range of fluids than water and steam.

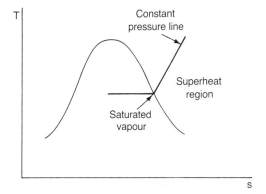

Figure 4.5 Variation of temperature in the superheat region.

Rogers and Mayhew (1988) gives a series of tables for not only steam, but also ammonia, Refrigerant-12, mercury vapour and several gases. For convenience, a shortened version of the property tables for the widely used thermodynamic fluids – water/steam, Refrigerant-12 and Refrigerant-134a (a more 'environmentally friendly' replacement for Refrigerant-12), are given at the end of the book in Appendix A. The properties quoted are the fluid pressure P, saturation temperature T_s, specific volume v, specific internal energy u, specific enthalpy h, and specific entropy s.

When considering the properties of a wet vapour it is cumbersome to continually refer to either saturated vapour or saturated liquid. The accepted notation is to define the properties of a saturated liquid by the subscript 'f' and those for a saturated vapour by the subscript 'g'. Consequently, the thermal properties for a saturated liquid are

$$v_f, \quad u_f, \quad h_f, \quad s_f$$

and those for a saturated vapour are

$$v_g, \quad u_g, \quad h_g, \quad s_g$$

Numeric values of these properties are defined in Appendix A for particular pressures. Values of the properties for steam at pressures other than those quoted, can be found in Rogers and Mayhew (1988).

The only marked difference between the properties tabulated in Rogers and Mayhew (1988) and those given in Appendix A, is that the former quotes pressures in bar, i.e. 100 kPa, whereas the pressures given in Appendix A are quoted in kPa.

Using the dryness fraction x together with the saturated properties, the properties of a wet vapour can be defined. The dryness fraction can be defined in terms of the mass of the saturated liquid m_f and the mass of the saturated vapour m_g:

$$x = \frac{m_g}{m_f + m_g} \tag{4.1}$$

For 1 kg of wet vapour, the mass of the saturated vapour becomes

$$m_g = x$$

and the mass of the saturated liquid

$$m_f = 1 - x$$

As explained in example 4.2 the specific volume of a liquid is negligible compared to that for a saturated vapour. Therefore, the specific volume for a wet vapour can be estimated from

$$v = x \times v_g \tag{4.2}$$

The specific internal energy of a wet vapour can be found from the sum of the internal energy of the saturated liquid together with the internal energy of the saturated vapour:

$$u = (1 - x)u_f + x \times u_g$$

Re-arranging gives

$$u = u_f + x (u_g - u_f) \tag{4.3}$$

(It should be noted that values of u_g and u_f are quoted in Appendix A1. Some textbooks and tables of properties give the difference $(u_g - u_f)$ as u_{fg}, with appropriate values.)

Similarly, the specific enthalpy and specific entropy for a wet vapour are given by the relationships

$$h = h_f + x (h_g - h_f) \tag{4.4}$$

$$s = s_f + x (s_g - s_f) \tag{4.5}$$

Example 4.3
One kg of wet steam has a volume of 0.1 m³ at a pressure of 200 kPa. If the volume remains constant and heat is added until the pressure reaches 800 kPa, find the initial and final dryness fraction of the steam.

Conceptual model

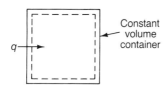

Process diagram

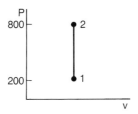

Analysis – from the information provided,

$$v_2 = v_1 = 0.1 \text{ m}^3/\text{kg}$$

From Appendix A1, at 200 kPa, $v_g = 0.8856$ m³/kg.
From equation (4.2)

$$x_1 = v_1 / v_g = \frac{0.1}{0.8856} = 0.113$$

Similarly, at 800 kPa, $v_g = 0.2403$ m³/kg and from equation (4.2)

$$x_2 = v_2 / v_g = \frac{0.1}{0.2403} = 0.416$$

Example 4.4

Calculate the heat input for the process defined in example 4.3.

Conceptual model – as for example 4.3.

Process diagram – as for example 4.3.

Analysis – applying equation (3.4) for a closed system

$$q - w = \Delta u$$

but, for constant volume, $w = 0$, hence

$$q = u_2 - u_1$$

Taking values from Appendix A1:

at 200 kPa, $u_1 = u_f + x_1 (u_g - u_f)$
$$= 505 + 0.113 (2530 - 505)$$
$$= 733.8 \text{ kJ/kg}$$
at 800 kPa, $u_2 = u_f + x_2 (u_g - u_f)$
$$= 720 + 0.416 (2577 - 720)$$
$$= 1492.5 \text{ kJ/kg}$$
heat input, $q = 1492.5 - 733.8$
$$= 758.7 \text{kJ/kg}$$

4.4.1 Interpolating between tabulated values

The tabulated values for wet steam given in Appendix A1 are only provided for a limited range of pressures. Where values are required at pressures other than those quoted, a more detailed range of values is given in Rogers and Mayhew (1988).

In the case of superheated steam values given in Appendix A2, the range of superheated temperatures is also limited. For example, it might be necessary to find the enthalpy h for superheated steam at a pressure of 200 kPa and a temperature of 220°C. Values of enthalpy are quoted for this pressure at temperatures of both 200°C and 250°C, but not at 220°C. In order to find a value of 220°C it can be assumed that properties vary directly with the temperature and that it is sufficiently accurate to interpolate between those values quoted for 200°C and 250°C.

To illustrate interpolation it is easier to use a worked example, as given in example 4.5.

Example 4.5

One kg of superheated steam, at 200 kPa and 220°C, is contained within a cylinder and frictionless piston. Heat is rejected from the system until the steam goes through a constant pressure process down to a dryness fraction of 0.5. Estimate the heat rejected from the system.

Conceptual model

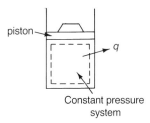

Process diagram

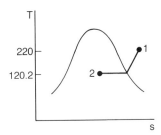

Analysis – applying equation (3.9) for a constant pressure process

$$q = h_2 - h_1$$

At state 1, the value of h_1 can be found by interpolation between 200°C and 250°C. From Appendix A2:

h at 200°C = 2871 kJ/kg
h at 250°C = 2971 kJ/kg

Interpolating between these values,

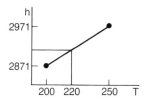

$$h_1 \text{ at } 220°C = 2871 + \frac{20}{50}(2971 - 2871)$$

$$= 2911 \text{ kJ/kg}$$

At state 2, taking values from Appendix A1,

$$
\begin{aligned}
h_2 &= h_f + x_2 (h_g - h_f) \\
&= 505 + 0.5\,(2707 - 505) \\
&= 1606 \text{ kJ/kg}
\end{aligned}
$$

Therefore, the heat transferred

$$q = h_2 - h_1 = 1606 - 2911$$
$$= -1305 \text{ kJ/kg}$$

Note – the negative sign indicates heat transfer **from** the system **to** the surroundings.

4.5 PERFECT GASES

Gases are highly superheated vapours and under normal operating conditions they exist at temperatures well in excess of the saturation temperature. Under these conditions gases tend to obey Boyle's law and Charles' law.

Boyle's law states that at constant temperature the variation of pressure with volume follows the relationship

$$P \times v = \text{constant}$$

Charles' law states that at constant pressure the variation of volume with temperature follows the relationship

$$\frac{v}{T} = \text{constant}$$

Combining these gives the equation

$$\frac{Pv}{T} = \text{constant}$$

The constant is given the symbol R and has the units J/kg K. It is generally referred to as the 'gas constant'. More correctly it should be called the 'specific gas constant' as it applies to one unit mass of the gas. Using the gas constant the equation is usually presented in the form

$$Pv = RT \tag{4.6}$$

Equation (4.6) is called the equation of state for a perfect gas. Although no real gas obeys this equation absolutely, it can be used to approximate real gas behaviour. In order to simplify the analysis of gases, the concept of a 'perfect gas' is used. A perfect gas, sometimes referred to as an ideal gas, is defined as one that does obey the equation of state, equation (4.6). From this definition it can be assumed that real gases can be considered to behave like a perfect gas in order to analyse thermodynamic situations.

The gas constant R has a unique value for each individual gas. For example, air has a value of 0.287 kJ/kg K, whereas oxygen has a value of 0.260 kJ/kg K. The uniqueness of each value of R stems from the fact that each unit mass, 1 kg, of gas contains a **different** number of molecules. This number depends on the molecular weight of the gas.

In section 1.11, a molecular mass was defined, called the 'kmol'. The actual mass in one kmol of gas can be determined by using the molecular weight of the gas. From equation (1.5), the mass in 1 kmol is given by

$$m = M \tag{4.7}$$

where m is the mass of the gas in kg and M is the value of the molecular weight of the gas. Using this relationship it is possible to use the molar mass as a basis for the gas constant instead of the actual mass.

Using the kmol as the basis, the gas constant is found to have a value of 8.314 kJ/kmol K, irrespective of the gas being considered. This value is called the 'universal gas constant' and is given the symbol R_o. The gas constant for any particular gas can be found from the universal gas constant, by applying equation (4.7)

$$R = \frac{R_o}{m} = \frac{R_o}{M} \qquad (4.8)$$

Example 4.6
Find the specific volume of nitrogen, $M = 28$, at a pressure of 200 kPa and a temperature of 15°C.

Analysis – applying equation (4.8)

$$R = \frac{R_o}{M} = \frac{8.314}{28} = 0.297 \text{ kJ/kg K}$$

Using this value in the equation of state equation (4.6),

$$v = \frac{RT}{p} = \frac{0.297 \times 10^3 \times 288}{200 \times 10^3} = 0.428 \text{ m}^3/\text{kg}$$

Example 4.7
A room 5 m × 4 m × 3 m contains 72 kg of air at a pressure of 101 kPa. Find the temperature of the air in the room assuming $R = 0.287$ kJ/kg K.

Analysis – the volume of the room $= 5 \times 4 \times 3 = 60$ m³. Hence the specific volume:

$$v = \frac{60}{m} = \frac{60}{72} = 0\,833 \text{ m}^3/\text{kg}$$

Applying the equation of state,

$$T = \frac{Pv}{R} = \frac{101 \times 10^3 \times 0.833}{0.287 \times 10^3} = 293.1 \text{ K} \quad (\text{i.e. } 20.1°C)$$

4.6   SPECIFIC HEATS OF GASES

The specific heat of a substance is defined as the heat required to raise one unit mass of that substance through a temperature rise of one degree. In the SI system the specific heat has units of J/kg K. Solids and liquids are considered to be incompressible, as the specific volume does not change significantly, and the specific heat for such substances is taken as a constant.

In the case of a gas, there are many processes that can take place. The previous chapter discussed four such processes:

constant volume
constant pressure
isothermal
adiabatic

Looking at each of these in turn the only two that result in a change in temperature of a perfect gas inside a closed system, due to heat transfer from the surroundings, are the constant volume and constant pressure processes. The isothermal process takes place without any change of temperature and the adiabatic process has zero heat transfer from the surroundings. Therefore, two specific heats for gases that need to be defined are:

specific heat at constant volume C_v;
specific heat at constant pressure C_P.

A more rigorous discussion of these properties is given by Look and Sauer (1988). However, within the context of this book it can be assumed that, for a perfect gas, the values of C_v and C_P are constant for any given gas at all values of pressure and temperature.

For a reversible constant volume process the heat transfer to, or from, a perfect gas is given by the equation

$$q = C_v (T_2 - T_1) \tag{4.9}$$

Similarly, for a reversible constant pressure process the heat transfer to, or from, a perfect gas is given by the equation

$$q = C_P (T_2 - T_1) \tag{4.10}$$

The stipulation that equations (4.9) and (4.10) are only true for reversible processes is necessary. Any friction within a constant pressure process or any turbulent motion of the gas during a constant volume process would result in discrepancies between the heat transferred and the resulting temperature changes.

4.6.1 Constant volume process for a perfect gas

Consider a closed system having a fixed boundary and containing 1 kg of a perfect gas, as shown in Figure 4.6. Because the boundary is fixed the volume of the closed system is constant throughout the process there can be no work crossing the boundary. Applying the first law of thermodynamics for a closed system, equation (3.3) with zero work, gives

$$q = u_2 - u_1$$

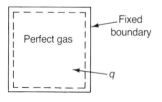

Figure 4.6 A closed system having constant volume.

The heat transferred from the surroundings takes place at constant volume. Therefore, from equation (4.9),

$$C_v (T_2 - T_1) = u_2 - u_1 \tag{4.11}$$

For a small change of temperature ΔT there will be a corresponding small change in the specific internal energy Δu, and the specific heat at constant volume can be defined as

$$C_v = \frac{\Delta u}{\Delta T} \qquad\qquad (4.12)$$

The value of C_v for a perfect gas is assumed to be constant, so that the change of specific internal energy is a function of temperature alone:

$$u = f(T)$$

Hence, the specific internal energy of a perfect gas varies linearly with absolute temperature and the value is taken as zero at absolute zero:

$$u = 0 \text{ at } T = 0$$

Example 4.8

A closed system, having a constant volume, contains 1 kg of a perfect gas. During a reversible process 75 kJ of heat is transferred to the system resulting in the temperature of the gas increasing from 20°C to 120°C. Calculate the change in internal energy, the value of C_v and the ratio of the final pressure to the initial pressure.

Conceptual model

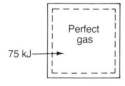

Process diagram

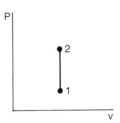

Analysis – equation (3.3)

$$q - w = u_2 - u_1$$

Since the volume is constant, $w = 0$. Therefore,

$$q = u_2 - u_1$$
$$q = 75 \text{ kJ/kg}$$

The change of internal energy $(u_2 - u_1) = 75$ kJ/kg
Taking equation (4.12)

$$C_v = \frac{\Delta u}{\Delta T}$$

where $\Delta T = 120 - 20 = 100$ K

$$C_V = \frac{75}{100} = 0.75 \text{ kJ/kg K}$$

From the equation of state, (4.6)

$$\frac{P}{T} = \frac{R}{v} = \text{constant}$$

For a process

$$\frac{P_1}{P_1} = \frac{P_2}{T_2}$$

Therefore,

$$\frac{P_2}{P_1} = \frac{T_2}{T_1} = \frac{(120 + 273)}{(20 + 273)} = 1.34$$

4.6.2 Constant pressure process for a perfect gas

Consider a closed system containing 1 kg of a perfect gas and undergoing a reversible constant pressure process, as shown in Figure 4.7.

From the discussion in section 3.7, it was shown that during a constant pressure process, the heat transferred to the system is equal to the change of specific enthalpy of the fluid within the system. From equation (3.9)

$$q = h_2 - h_1$$

The heat transferred from the surroundings takes place at constant pressure, equation (4.10):

$$q = C_P (T_2 - T_1)$$

Therefore,

$$q = C_P (T_2 - T_1) = h_2 - h_1 \tag{4.13}$$

For a small change of temperature ΔT there will be a corresponding small change in the specific enthalpy Δh, and the specific heat at constant pressure can be defined as

$$C_P = \frac{\Delta h}{\Delta T} \tag{4.14}$$

It follows that $h = f(T)$. Since $h = u + Pv$ and both u and P are zero at absolute zero, then

$$h = 0 \text{ at } T = 0$$

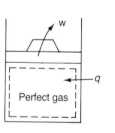

Figure 4.7 A closed system under constant pressure.

Example 4.9

A closed system, consisting of a cylinder and frictionless piston, contains 1 kg of a perfect gas having a molecular weight of 26. The piston is loaded so that the pressure in the system is constant at 200 kPa. The system undergoes a

process such that the volume changes from 0.5 to 1 m³. Find the heat transferred during this process assuming that $C_P = 1.08$ kJ/kg K.

Conceptual model

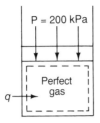

Process diagram

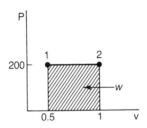

Analysis – from equations (3.9) and (4.13)

$$q = h_2 - h_1 = C_P (T_2 - T_1)$$

It is possible to calculate T_1 and T_2 by applying the equation of state, (4.6):

$$Pv = RT$$

From equation (4.8)

$$R = \frac{R_o}{M} = \frac{8.314}{26} = 0.32 \text{ kJ/kg K}$$

Substituting in equation (4.6)

$$T_1 = \frac{P_1 v_1}{R} = \frac{200 \times 10^3 \times 0.5}{0.32 \times 10^3} = 312.5 \text{ K}$$

$$T_2 = \frac{P_2 v_2}{R} = \frac{200 \times 10^3 \times 1}{0.32 \times 10^3} = 625 \text{ K}$$

Therefore,

$$q = C_P (T_2 - T_1)$$
$$= 1.08 (625 - 312.5) = 337.5 \text{ kJ/kg}$$

4.6.3 Relationship between the specific heats

The definition of specific enthalpy is

$$h = u + Pv$$

For a small change of specific enthalpy there will be a corresponding change in the specific internal energy and the work done at constant pressure:

$$\Delta h = \Delta u + P\Delta v \qquad (4.15)$$

From the definitions of C_v and C_P, equations (4.12) and (4.14), for a perfect gas

$$\Delta u = C_v\Delta T$$
$$\Delta h = C_P\Delta T$$

From the equation of state (4.6) for a perfect gas the work done can be expressed as

$$P\Delta v = R\Delta T$$

Combining these in equation (4.15)

$$C_P \Delta T = C_v \Delta T + R\Delta T$$

Cancelling out the change of temperature ΔT gives a relationship between the specific heats for a perfect gas:

$$C_P = C_v + R \qquad (4.16)$$

Example 4.10

When 1 kg of a perfect gas is heated from 20°C to 80°C at constant pressure, the heat input is 120 kJ. When the same gas is heated at constant volume, the heat input is 90 kJ. Find the gas constant for the perfect gas.

Analysis – for the constant pressure process, using equation (4.10),

$$C_P = \frac{q}{T_2 - T_1} = \frac{120}{60} = 2\,\text{kJ/kg K}$$

For the constant volume process, from equation (4.9),

$$C_v = \frac{q}{T_2 - T_1} = \frac{90}{60} = 1.5\,\text{kJ/kg K}$$

From equation (4.16)

$$R = C_v - C_P = 2 - 1.5 = 0.5\,\text{kJ/kg K}$$

4.7   ISOTHERMAL PROCESS FOR A PERFECT GAS

During an isothermal process the temperature of the perfect gas remains constant. Substituting this condition in the equation of state, (4.6), the process obeys the law:

$$Pv = \text{constant}$$

Figure 4.8(a) shows a closed system, consisting of a cylinder and frictionless piston, in which 1 kg of a perfect gas undergoes a reversible isothermal expansion. Figure 4.8(b) shows the process diagram for the expansion. During the

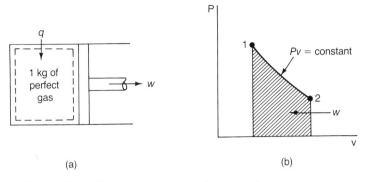

Figure 4.8 A closed system undergoing an isothermal expansion.

process, work is done on the surroundings and this can be evaluated from the area under the curve on the P–v diagram. From equation (3.6)

$$w = \int_1^2 P \, dv$$

but

$$P = \text{constant}/v$$

so that

$$w = \text{constant} \int_1^2 \frac{dv}{v}$$

$$= \text{constant} \ln \frac{v_2}{v_1}$$

At state 1

$$P_1 v_1 = Pv = \text{constant}$$

so that the work is given by

$$w = P_1 v_1 \ln \frac{v_2}{v_1} \tag{4.17}$$

This value of work can be substituted in the first law of thermodynamics, equation (3.4):

$$q - w = u_2 - u_1$$

However, the change in specific internal energy, $u_2 - u_1$, is proportional to the change of temperature. Since there is no change of temperature, it follows that

$$u_2 - u_1 = 0$$

and the heat transfer is equal to the work done:

$$q = w \tag{4.18}$$

Therefore, the heat transferred during the reversible isothermal process of a perfect gas, can also be evaluated from equation (4.17).

Example 4.11

A perfect gas undergoes a reversible isothermal compression from a pressure of 100 kPa and a temperature of 20°C, to a pressure of 500 kPa. Find the work done during the process for 1 kg of the gas. Assume the gas has a molecular weight of 28.

Process diagram

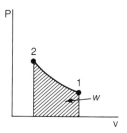

Analysis – the work done for an isothermal compression can be found from equation (4.17),

$$w = P_1 v_1 \ln \frac{v_2}{v_1}$$

Of the variables, v_1 and the ratio v_2/v_1 must be evaluated.
From the equation of state, (4.6)

$$v_1 = \frac{RT_1}{P_1}$$

and R can be evaluated from equation (4.8) as

$$R = \frac{R_o}{M} = \frac{8.314}{28} = 0.297 \text{ kJ/kg K}$$

Hence

$$v_1 = \frac{0.297 \times 10^3 \times 293}{100 \times 10^3} = 0.87 \text{ m}^3/\text{kg}$$

For an isothermal process

$$Pv = \text{constant}$$

so that

$$P_1 v_1 = P_2 v_2$$

and

$$\frac{v_2}{v_1} = \frac{P_1}{P_2} = \frac{100}{500} = 0.2$$

Substituting in equation (4.17)

$$w = 100 \times 10^3 \times 0.87 \times \ln 0.2 = -140 \times 10^3 \text{ J} = -140 \text{ kJ}$$

Note – the negative sign indicates that the work is done on the system during a compression process.

4.8   ADIABATIC PROCESS FOR A PERFECT GAS

During an adiabatic process for a perfect gas, the heat transfer to or from the system is zero. The change of pressure with respect to the volume obeys the law

$$Pv^\gamma = \text{constant} \qquad (4.19)$$

where the value of the index γ is defined below.

Figure 4.9(a) shows a closed system, consisting of an insulated cylinder and frictionless piston, in which 1 kg of a perfect gas undergoes a reversible adiabatic expansion. Figure 4.9(b) shows the process diagram for the expansion. During the process, work is done on the surroundings and this can be evaluated from the area under the curve on the P–v diagram, from equation (3.6):

$$w = \int_1^2 P/\mathrm{d}v$$

but

$$P = \text{constant}/v^\gamma$$

so that

$$w = \text{constant} \int_1^2 \frac{\mathrm{d}v}{v^\gamma}$$

$$= \text{constant} \left[\frac{v^{1-\gamma}}{1-\gamma} \right]_1^2$$

At state 1

$$P_1 v_1^\gamma = \text{constant}$$

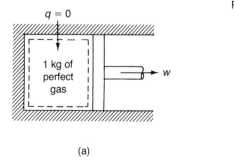

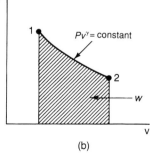

(a)   (b)

Figure 4.9 A closed system undergoing an adiabatic expansion.

and at state 2

$$P_2 v_2^{\gamma} = \text{constant}$$

Substituting gives

$$w = \frac{P_2 v_2 - P_1 v_1}{1 - \gamma} \tag{4.20}$$

This value of work can be substituted in the first law of thermodynamics, equation (3.4):

$$q - w = u_2 - u_1$$

The heat transferred q is zero, so that substituting for w by equation (4.20) gives

$$-\left(\frac{P_2 v_2 - P_1 v_1}{1 - \gamma}\right) = u_2 - u_1 \tag{4.21}$$

From the equation of state, (4.6)

$$P_2 v_2 = R T_2 \quad \text{and} \quad P_1 v_1 = R T_1$$

and, from equation (4.11),

$$u_2 - u_1 = C_v (T_2 - T_1)$$

Substituting these values into equation (4.21) yields

$$-\frac{R}{1 - \gamma} (T_2 - T_1) = C_v (T_2 - T_1)$$

so that for a reversible adiabatic process

$$\frac{R}{\gamma - 1} = C_v$$

and this can only be true if the index γ has the value

$$\gamma = \frac{C_P}{C_v} \tag{4.22}$$

Using this definition of γ, it is possible to derive a series of relationships between the properties of a perfect gas during an adiabatic process.

The relationship between pressure and volume is given by equation (4.19) and can be expressed as

$$P_1 v_1^{\gamma} = P_2 v_2^{\gamma} \tag{4.23}$$

From the equation of state

$$P_1 = \frac{R T_1}{v_1} \quad \text{and} \quad P_2 = \frac{R T_2}{v_2}$$

Substituting these values in equation (4.23) gives a relationship between temperature and volume:

$$T_1 v_1^{\gamma - 1} = T_2 v_2^{\gamma - 1}$$

or

$$\frac{T_1}{T_2} = \left(\frac{v_2}{v_1}\right)^{\gamma-1} \tag{4.24}$$

From the equation of state

$$v_1 = \frac{RT_1}{P_1} \quad \text{and} \quad v_2 = \frac{RT_2}{P_2}$$

Substituting these values in equation (4.23) gives a relationship between temperature and pressure:

$$P_1\left(\frac{RT_1}{P_1}\right)^{\gamma} = P_2\left(\frac{RT_2}{P_2}\right)^{\gamma}$$

therefore

$$\frac{T_1^{\gamma}}{P_1^{\gamma-1}} = \frac{T_2^{\gamma}}{P_2^{\gamma-1}}$$

or

$$\frac{T_1}{T_2} = \left(\frac{P_1}{P_2}\right)^{(\gamma-1)/\gamma} \tag{4.25}$$

Example 4.12

Air undergoes a reversible adiabatic expansion from a pressure of 500 kPa and a temperature of 800 K, down to a pressure of 200 kPa. Assuming that air behaves as a perfect gas with values of $C_P = 1.005$ kJ/kg K and $R = 0.287$ kJ/kg K, calculate the temperature of the air at the end of the expansion process.

Conceptual diagram – as shown in Figure 4.9(a)

Process diagram – since the process is related to the variation of temperature with pressure, the usual P–v or T–s diagrams have little value in this problem.

Analysis – the variation of temperature with pressure is governed by equation (4.25) for a reversible adiabatic process. However, it is necessary to evaluate the value of γ. From equation (4.22)

$$\gamma = \frac{C_P}{C_v}$$

and from equation (4.16)

$$C_v = C_P - R = 1.005 - 0.287 = 0.718 \text{ kJ/kg K}$$

Therefore,

$$\gamma = \frac{1.005}{0.718} = 1.4$$

Substituting the values in equation (4.25)

$$\frac{T_1}{T_2} = \left(\frac{500}{200}\right)^{\frac{1.4-1}{1.4}} = 1.299$$

and

$$T_2 = \frac{T_1}{1.299} = \frac{800}{1.299} = 615.7\ \text{K}$$

4.9   REAL GASES

The foregoing discussion of the properties of gases is based upon the assumption that the behaviour of a gas approximates to that of a perfect gas. A perfect gas is defined as one that is governed by the following laws:

1. the equation of state, $Pv = RT$;
2. the values of specific heat, C_v and C_P, remain constant.

In practice, a real gas does **not** obey **either** of these. For example, Figure 4.10 shows the variation of C_P for air over a temperature range 300–1000 K. At moderate temperatures, the variation is slight, but above 500 K the value of C_P increases quite rapidly, although the rate of increase is reduced above 1000 K.

Fortunately, the errors incurred in using the equation of state, $Pv = RT$, are not so significant. It is possible to analyse engines, or thermodynamic situations, in a realistic manner by assuming that this equation of state remains valid and that the specific heat value only varies with temperature. This assumes that the gas behaves as, what is called, a 'semi-perfect' gas. To distinguish between the characteristics of a perfect, semi-perfect and real gas, the criteria are tabulated on p. 85 with a tick, signifying the type of gas that complies with each criterion.

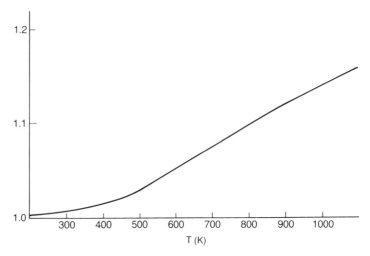

Figure 4.10 Variation of C_P with temperature for air.

Characteristic	Perfect	Semi-perfect	Real
$Pv = RT$	✓	✓	✗
Constant C_P	✓	✗	✗

However, the analysis of a semi-perfect, or real, gas is outside the scope of this introductory text. Look and Sauer (1988) gives a more detailed discussion of this topic.

SUMMARY

In this chapter the properties of liquids, vapours and perfect gases have been discussed. The key terms that have been introduced are:

two-phase fluid
saturation temperature
saturated liquid
saturated vapour
wet liquid
dryness fraction
critical point
superheated vapour
perfect gas
semi-perfect gas
real gas

Key equations that have been introduced.

For a wet vapour:

$$x = \frac{m_g}{m_f + m_g} \tag{4.1}$$

$$v = xv_g \tag{4.2}$$

$$u = u_f + x\,(u_g - u_f) \tag{4.3}$$

$$h = h_f + x\,(h_g - h_f) \tag{4.4}$$

$$s = s_f + x\,(s_g - s_f) \tag{4.5}$$

For a perfect gas:

$$Pv = RT \tag{4.6}$$

$$R = \frac{R_0}{M} \tag{4.8}$$

For a perfect gas undergoing a constant volume process:

$$q = C_v\,(T_2 - T_1) \tag{4.9}$$

where

$$C_v = \frac{\Delta u}{\Delta T} \tag{4.12}$$

For a perfect gas undergoing a constant pressure process:

$$q = C_P (T_2 - T_1) \tag{4.10}$$
$$C_P (T_2 - T_1) = h_2 - h_1 \tag{4.13}$$

where

$$C_P = \frac{\Delta h}{\Delta T} \tag{4.14}$$

Relationship between the specific heats for a perfect gas:

$$C_P = C_v + R \tag{4.16}$$

$$\gamma = \frac{C_P}{C_v} \tag{4.22}$$

For a perfect gas undergoing an isothermal process:

$$w = P_1 v_1 \ln \frac{v_2}{v_1} \tag{4.17}$$

For a perfect gas undergoing an adiabatic process:

$$P_1 v_1^\gamma = P_2 v_2^\gamma \tag{4.23}$$

$$\frac{T_1}{T_2} = \left(\frac{v_2}{v_1} \right)^{\gamma - 1} \tag{4.24}$$

$$\frac{T_1}{T_2} = \left(\frac{P_1}{P_2} \right)^{\gamma - 1/\gamma} \tag{4.25}$$

PROBLEMS

Two-phase fluids

1. For steam at a pressure of 1 MPa find:
   (a) the specific internal energy if $x = 0.75$
   (b) the dryness fraction if $h = 2000 \, kJ/kg$
   (c) the temperature if $s = 7 \, kJ/kg \, K$.

2. One kg of Refrigerant-12 is contained in a closed rigid vessel. If the refrigerant is originally a saturated vapour at 30°C and is cooled to a pressure of 308.6 kPa, determine the final dryness fraction.

3. A mass of steam at a pressure of 400 kPa and a dryness fraction of 0.6 is contained within a cylinder and frictionless piston. Heat is transferred to the system so that the steam goes through a reversible constant pressure process to a temperature of 280°C. Estimate the specific heat transfer to the system.

4. Within a cylinder and frictionless piston a quantity of Refrigerant-134a is raised from a dryness fraction of 0.5 to 0.85 at a constant temperature of

20°C. Assuming the process to be reversible, find the heat transfer to the system for each kg of refrigerant.

5. In an insulated closed system, one kg of steam expands from 2 MPa in a saturated vapour condition, down to a pressure of 200 kPa and dryness fraction of 0.7. Assuming the expansion to be reversible, calculate the specific work done.

Perfect gases

6. A closed vessel contains carbon dioxide, $M = 44$, at a pressure of 200 kPa and a temperature of 20°C. If the vessel has a fixed volume of 0.5 m³ find the mass of gas contained.

7. A cylinder and frictionless piston contains 1 kg of a perfect gas. During a reversible constant pressure process, 85 kJ of heat is transferred to the system resulting in the temperature of the gas increasing from 50°C to 150°C. Calculate the change in specific enthalpy, the value of C_P and the ratio of the final volume to the initial volume.

8. A perfect gas undergoes a reversible isothermal expansion from a pressure of 800 kPa and a temperature of 100°C, down to a pressure of 200 kPa. Find the heat transferred during the process for 1 kg of the gas. Assume the gas has a molecular weight of 32.

9. Air undergoes a reversible adiabatic compression from atmospheric conditions of 100 kPa and 20°C, to a pressure of 500 kPa. Assuming that air behaves as a perfect gas with values of $C_v = 0.718$ kJ/kg K and $R = 0.287$ kJ/kg K, calculate the temperature at the end of the process and the work done.

5 Flow processes

5.1 AIMS

- To introduce the mass flow rate for steady flow through an open system.
- To define the flow work for an open system.
- To introduce the steady flow energy equation for an open system.
- To apply the SFEE to analyse steady flow in a constant pressure device.
- To apply the SFEE to analyse steady flow in an adiabatic device with no work.
- To apply the SFEE to analyse steady flow in an adiabatic device with work.
- To introduce the steady flow energy equation as a rate equation.
- To apply the SFEE as a rate equation to the analysis of a heat exchanger.

5.2 STEADY FLOW THROUGH AN OPEN SYSTEM

An open system has been defined in Chapter 2, as one in which the volume inside the boundary is fixed but for which fluid may flow into or out of the system across the boundary. Figure 5.1 shows a typical open system.

Across the boundary there can be transfer of energy in the form of both heat and work. In addition, there is an input of fluid entering at 1 and an output of fluid leaving at 2. The fluid flow can be measured in terms of the mass flow rate, the mass of fluid entering or leaving in a unit time, i.e. $m^\cdot$ given in units of kg/s. It is assumed that the mass flow rate of fluid leaving the open system equals the mass flow rate of fluid entering. This means that there is no storing of the fluid in the system.

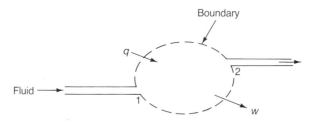

Figure 5.1 An open thermodynamic system.

If the mass flow rate remains constant and does not change with time, the flow is referred to as 'steady'. For steady flow as applied to the open system shown in Figure 5.1, there is conservation of mass:

$$\dot{m}'_1 = \dot{m}'_2 = \dot{m}'$$

The mass flow rate is equal to the volume flow rate, velocity × flow area, divided by the specific volume of the fluid:

$$\dot{m} = \frac{V_1 A_1}{v_1} = \frac{V_2 A_2}{v_2} \tag{5.1}$$

Example 5.1

Saturated steam at a pressure of 600 kPa enters an open thermodynamic system at a velocity of 100 m/s through a pipe of 0.2 m diameter. If the steam leaves the open system at a pressure of 200 kPa and a dryness fraction of 0.75, through an exit pipe of 0.5 m diameter, find:

1. the mass flow rate;
2. the velocity of the steam leaving the system.

Conceptual model

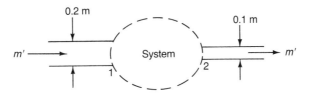

Analysis – from equation (5.1)

$$\dot{m} = \frac{V_1 A_1}{v_1}$$

and from Appendix A1

$$v_1 = 0.3156 \, \text{m}^3/\text{kg}$$

Also

$$A_1 = \frac{\pi}{4} d_1^2 = \frac{\pi}{4} (0.2)^2 = 0.0314 \, \text{m}^3$$

Therefore,

$$\dot{m} = \frac{100 \times 0.0314}{0.3156} = 9.95 \, \text{kg/s}$$

It follows that

$$9.95 = \frac{V_2 A_2}{v_2}$$

From equation (4.2)

$$v_2 = xv_g = 0.75 \times 0.8856$$
$$= 0.6642 \text{ m}^3/\text{kg}$$

Also

$$A_2 = \frac{\pi}{4}(0.5)^2 = 0.196 \text{ m}^3$$

Therefore,

$$9.95 = \frac{V_2 \times 0.196}{0.6642}$$

$$V_2 = 33.7 \text{ m/s}.$$

5.3 FLOW WORK

Consider the open system shown in Figure 5.1. The fluid entering the system at section 1 is being pushed in by the fluid behind it. Therefore, work is being done on the fluid inside the system. Similarly, the fluid leaving the system at section 2 has to push the fluid in front of it, representing work out of the system. The work done in moving the fluid is called the 'flow work'.

Figure 5.2 shows a simplified system in which fluid is flowing along a pipe. At section (a) the pressure on the fluid is P and this causes the fluid to flow to the right. Taking 1 kg of the fluid, the flow of this quantity through section (a) moves the fluid to section (b), a distance of 'y'. The flow work is given by

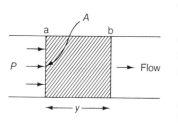

Figure 5.2 Flow along a pipe.

$$\text{flow work} = \text{force} \times \text{distance}$$
$$= (P \times A) \times y$$
$$= P(A \times y)$$

Since the volume defined by $(A \times y)$ contains 1 kg of fluid, this volume represents the specific volume v. Hence, the flow work can be expressed as

$$\text{flow work} = P\,v \tag{5.2}$$

For the open system shown in Figure 5.1 the flow work being done on the fluid in the system is P_1v_1 while the flow work being done on the fluid leaving the system is P_2v_2. Therefore, the net flow work for the system is given by

$$\text{net flow work} = P_1v_1 - P_2v_2 \tag{5.3}$$

Example 5.2
Water enters a pump at a pressure of 100 kPa and leaves at a pressure of 500 kPa. Calculate the specific work done, assuming that water has a constant specific volume of 0.001 m³/kg.

Conceptual model

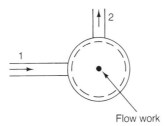

Flow work

Process diagram

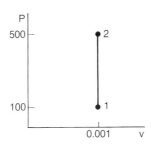

Analysis – from equation (5.3)

$$\text{net flow work} = P_1 v_1 - P_2 v_2$$

and, since $v_1 = v_2$,

$$\text{net flow work} = v\,(P_1 - P_2)$$
$$= 0.001\,(100 \times 10^3 - 500 \times 10^3)$$
$$= -400\ \text{J/kg}$$

Note – the negative sign for the work implies a work input to the pump. A study of the P–v diagram for the pump indicates that the area under the curve 1–2,

$$\int P\mathrm{d}v$$

is zero. In Chapter 3 it was shown that the work was equal to the area under the curve on the P–v diagram. The apparent contradiction between these two statements is due to the flow work being applicable to an open system, whilst the work for a process as represented by the area on the P–v diagram is true for a closed system.

5.4 STEADY FLOW ENERGY EQUATION

Having defined the flow work in the preceding section, it is now possible to evaluate the energy of the fluid entering and leaving an open system.

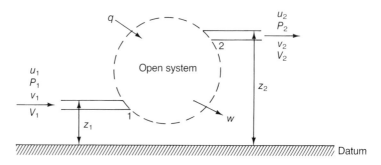

Figure 5.3 Flow characteristics across an open system.

Consider the fluid entering the open system shown in Figure 5.3 at section 1. Each kg of fluid enters with a particular value of internal energy u_1, kinetic energy $V_1^2/2$, potential energy $z_1 g$, and having flow work $P_1 v_1$ done on it. Combining these terms gives a total energy for the fluid entering at section 1 as

$$u_1 + P_1 v_1 + \frac{V_1^2}{2} + z_1 g$$

Similarly, the total energy of 1 kg of fluid leaving at section 2 is

$$u_2 + P_2 v_2 + \frac{V_2^2}{2} + z_2 g$$

The difference between these energy levels is given by the energy crossing the boundary of the system in the form of heat q and work w.

The first law of thermodynamics for an open system can, therefore, be expressed as

$$q - w = \left(u_1 + P_1 v_1 + \frac{V_1^2}{2} + z_1 g \right) - \left(u_2 + P_2 v_2 + \frac{V_2^2}{2} + z_2 g \right) \qquad (5.4)$$

Equation (5.4) is somewhat cumbersome, but fortunately it can be simplified in order to analyse real thermodynamic devices operating as open systems. It is found in practice that the change in height, and the resulting change in potential energy, are negligible and the zg terms can be ignored. Also, from equation (3.9), the term $(u + pv)$ can be expressed as the enthalpy h.

Applying these changes to equation (5.4) and re-arranging the terms, gives

$$q + h_1 + \frac{V_1^2}{2} = h_2 + \frac{V_2^2}{2} + w \qquad (5.5)$$

This is called the steady flow energy equation, SFEE for short.

To be consistent, the units of each of the terms in the steady flow energy equation must be the same. Taking the $V^2/2$ term, the units are

$$\frac{m^2}{s^2}$$

multiplying top and bottom by 'kg' gives

$$\frac{\text{kg m}^2}{\text{kg s}^2} = \frac{\text{kg m}}{\text{s}^2}\frac{\text{m}}{\text{kg}} = \frac{\text{N m}}{\text{kg}} = \text{J/kg}$$

It follows that q, w and h must be expressed in the same units.

Example 5.3

A steady flow device is insulated and operates with steam entering at a pressure of 600 kPa, a temperature of 250°C, and a velocity of 200 m/s. If the steam leaves the device as a saturated vapour at a pressure of 100 kPa, with a velocity of 50 m/s, calculate the specific work done by the device.

Conceptual model – as the device is insulated, assume $q = 0$.

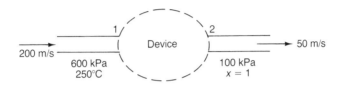

Analysis – applying the SFEE, equation (5.5)

$$q + h_1 + \frac{V_1^2}{2} = h_2 + \frac{V_2^2}{2} + w$$

since $q = 0$

$$w = h_1 - h_2 + \frac{V_1^2}{2} - \frac{V_2^2}{2}$$

From Appendix A2:

$$h_1 = 2958 \text{ kJ/kg} = 2958 \times 10^3 \text{ J/kg}$$

From Appendix A1:

$$h_2 = 2675 \text{ kJ/kg} = 2675 \times 10^3 \text{ J/kg}$$

Therefore:

$$w = 2958 \times 10^3 - 2675 \times 10^3 + \frac{(200)^2}{2} - \frac{(50)^2}{2}$$

$$= 301\,750 \text{ J/kg}$$

$$= 301.75 \text{ kJ/kg}$$

5.5   STEADY FLOW THERMODYNAMIC DEVICES

There are a wide range of steady flow devices used in thermodynamic situations. Boilers, turbines, condensers and pumps are used in steam power plants.

Compressors, combustion chambers and turbines are used in gas turbine engines. Aircraft gas turbine engines utilize additional steady flow devices in the form of diffusers and nozzles. Compressors, condensers, throttle valves and evaporators (boilers) are used in refrigeration plant.

All of these devices operate with steady flow under normal operating conditions, and can be analysed using the steady flow energy equation. It is not necessary to consider each individual device in a unique situation. A turbine in a steam plant can be considered to operate on the same basic principles as a turbine in a gas turbine engine. Similarly, a condenser in a refrigeration plant operates on the same basic principles as a condenser in a steam plant.

However, in order to analyse these devices it is necessary to categorize them in terms of the flow process through the device.

1. **Constant pressure process** – boilers, condensers and combustion chambers all operate with constant pressure processes through the devices. In the case of boilers and condensers there is a change of phase of the fluid, whereas combustion chambers operate with gases. All three types of device can be analysed in the same way using the steady flow energy equation.
2. **Adiabatic process with no work** – diffusers and nozzles are devices that are used to bring about a change of kinetic energy. A diffuser is used to slow the flow down while a nozzle speeds the flow up. Neither device has a work input or output. During the analysis of the flow characteristics it can be assumed that there is no heat transfer between the device and the surroundings, implying that the process is adiabatic. The change in velocity is achieved by means of a change of pressure across the device.

   Another device that can be considered to operate with an adiabatic process, which has no work across the boundary and achieves a change of pressure, is a throttle valve. Such a device is used in a refrigeration plant simply to bring about a reduction in the fluid pressure.
3. **Adiabatic process with work** – turbines, compressors and pumps all operate with work crossing the system boundary. In the case of a turbine, its purpose is to produce work. In the case of either a compressor or a pump, the purpose is to do work on the fluid in order to bring about a change in pressure. A compressor is used when the fluid is either a vapour or a gas, a pump is used when the fluid is a liquid. If the devices within this category are considered to be insulated, there is no heat transfer to or from the surroundings, and the process is considered to be adiabatic.

The analysis of steady flow devices within these three categories are considered in greater detail in the following sections.

5.5.1 Steady flow constant pressure process

The representation of either a boiler or a condenser by means of a schematic diagram has already been introduced in Figure 2.2. The same representation can be used for the combustion chamber of a gas turbine engine as shown in Figure 5.4.

Figure 5.4(a) shows a typical combustion chamber. Air enters from a compressor, is heated by internal combustion of a fuel, and leaves as hot gas to

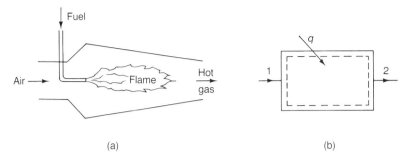

Figure 5.4 Schematic diagram of a combustion chamber.

drive a gas turbine. This situation can be modelled in a schematic diagram as shown in Figure 5.4(b). In reality, the mass flow rate of fuel is very much smaller than the mass flow rate of air. This will become clearer after the discussion of combustion in Chapter 11. For the present, it can be assumed that the mass flow rate of fuel is negligible compared to the mass flow rate of air and that, in Figure 5.4(b), the flow from 1 to 2 is both steady and equal. Although the air is heated by means of internal combustion, the energy is transferred into the combustion chamber in the form of a flow of fuel from outside. In Figure 5.4(b) this is modelled as heat addition q from the surroundings.

The open system shown in Figure 5.4(b) is not only typical of a combustion chamber, but also of a boiler and a condenser. Therefore, all three devices can be analysed in exactly the same way using the steady flow energy equation, (5.5)

$$q + h_1 + \frac{V_1^2}{2} = h_2 + \frac{V_2^2}{2} + w$$

This can be simplified by making the following assumptions:

1. there is no work done, so that $w = 0$;
2. the change of velocity across the device is small, so that $V_1^2 \approx V_2^2$.

Using these assumptions, the steady flow energy equation reduces to the form

$$q = h_2 - h_1 \qquad (5.6)$$

Example 5.4
A boiler operates at a pressure of 1 MPa. Feed water enters in a saturated condition and the steam leaves with a temperature of 300°C. Calculate the heat transfer to the boiler for each kg of water converted to superheated steam.

Conceptual model

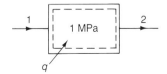

Process diagram

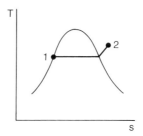

Analysis – the specific heat transfer is given by equation (5.6):

$$q = h_2 - h_1$$

From Appendix Al, the enthalpy of saturated water

$$h_1 = 763 \text{ kJ/kg}$$

From Appendix A2, the enthalpy of superheated steam

$$h_2 = 3052 \text{ kJ/kg}$$

Hence

$$q = 3052 - 763 = 2289 \text{ kJ/kg}$$

Example 5.5

Air enters the combustion chamber of a gas turbine engine at 600 K and leaves at a temperature of 1250 K. Assuming that the fuel has negligible effect on the heat transfer process and that air has a value of C_P of 1.005 kJ/kg K, calculate the heat input for each kg of air.

Conceptual model – as for example 5.4.

Process diagram – as the process is at constant pressure, it can be represented on the P–v diagram. However, the P–v diagram provides little insight into the heat transfer process. An alternative representation on the T–s diagram illustrates the effective increase in temperature during the process.

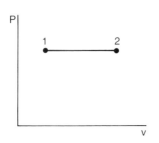
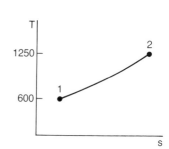

Analysis – the specific heat transfer is given by equation (5.6):

$$q = h_2 - h_1$$

For a perfect gas undergoing a constant pressure process, this can be expressed, using equation (4.10) as:

$$q = C_P (T_2 - T_1)$$

Therefore,

$$q = 1.005 \,(1250 - 600)$$
$$= 653.25 \text{ kJ/kg}$$

5.5.2 Steady adiabatic flow through a diffuser or nozzle

A diffuser or a nozzle can be represented by means of a schematic diagram as shown in Figure 5.5.

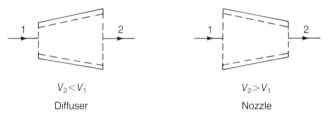

$V_2 < V_1$ ⠀⠀⠀⠀⠀⠀⠀⠀⠀ $V_2 > V_1$

Diffuser ⠀⠀⠀⠀⠀⠀⠀⠀⠀⠀ Nozzle

Figure 5.5 Schematic diagram of a diffuser and nozzle.

In principle, a diffuser is a device for decreasing flow velocity, while a nozzle is the reverse, used to increase the flow velocity. Both can be analysed in the same way using the steady flow energy equation, (5.5):

$$q + h_1 + \frac{V_1^2}{2} = h_2 + \frac{V_2^2}{2} + w$$

This can be simplified by making the following assumptions:

1. the flow is adiabatic, so there is no heat transfer across the boundary, and $q = 0$;
2. there is no work done, so that $w = 0$.

Using these assumptions, the steady flow energy equation reduces to the form

$$V_1^2 - V_2^2 = 2(h_2 - h_1) \tag{5.7}$$

In addition, it can be assumed that the friction is insignificant and that the flow process is reversible.

Example 5.6
The intake of a jet engine acts as a diffuser. Air enters at a pressure of 70 kPa, a temperature of –3°C and a velocity of 200 m/s. If the air leaves at a velocity of

100 m/s, find the exit temperature and pressure. Assume $C_P = 1.005$ kJ/kg K and $\gamma = 1.4$.

Conceptual model

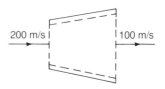

200 m/s | | 100 m/s

Process diagram – as the process is both adiabatic and reversible, it can be represented on the T–s diagram.

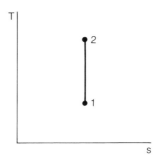

Analysis – apply equation (5.7):

$$V_1^2 - V_2^2 = 2(h_2 - h_1)$$

Since the fluid is a perfect gas, from equation (4.13),

$$h_2 - h_1 = C_P (T_2 - T_1)$$

and

$$V_1^2 - V_2^2 = 2 C_P (T_2 - T_1)$$

Substituting the values gives

$$(200)^2 - (100)^2 = 2 \times 1.005 \times 10^3 (T_2 - 270)$$
$$T_2 = 284.9 \text{ K}$$

For an adiabatic process, the relationship between temperature and pressure is given by equation (4.25):

$$\frac{T_1}{T_2} = \left(\frac{P_1}{P_2} \right)^{(\gamma - 1)/\gamma}$$

Re-arranging,

$$P_2 = P_1 \left(\frac{T_2}{T_1} \right)^{\gamma/(\gamma - 1)}$$

$$= 70 \left(\frac{284.9}{270} \right)^{1.4/0.4}$$

$$= 84.5 \text{ kPa}$$

Example 5.7

A steam turbine contains a stationary nozzle to increase the velocity of the steam into the rotating blades. If steam enters the nozzle at 800 kPa, 250°C, and leaves the nozzle at a pressure of 60 kPa, calculate the velocity of the steam at the nozzle exit. Assume the velocity of the steam entering the nozzle to be negligible.

Conceptual model

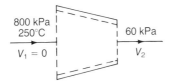

Process diagram – the steam enters in a superheated condition and leaves as wet steam. The process can be represented on the T–s diagram

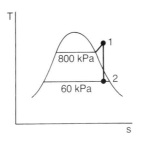

Analysis – apply equation (5.7):

$$V_1^2 - V_2^2 = 2(h_2 - h_1)$$

This can be re-arranged into the form

$$V_2^2 - V_1^2 = 2(h_1 - h_2)$$

but $V_1 = 0$, so that the velocity leaving is

$$V_2 = \sqrt{2(h_1 - h_2)}$$

At state 1, from Appendix A2,

$$h_1 = 2951 \text{ kJ/kg}$$
$$s_1 = 7.040 \text{ kJ/kg K}$$

As state 2

$$s_2 = s_1 = 7.040 \text{ kJ/kg K}$$

This can be used to find the dryness fraction x_2.
From equation (4.5):

$$s_2 = s_f + x_2 (s_g - s_f)$$

At state 2, taking properties from Appendix A1,

$$s_f = 1.145 \text{ kJ/kg K}, \quad s_g = 7.531 \text{ kJ/kg K}$$

Therefore,

$$7.040 = 1.145 + x_2 (7.531 - 1.145)$$
$$x_2 = 0.923$$

From equation (4.4)

$$h = h_f + x(h_g - h_f)$$

At state 2, taking properties from Appendix A1,

$$h_f = 360 \text{ kJ/kg}, \quad h_g = 2653 \text{ kJ/kg}$$

substituting,

$$h_2 = 360 + 0.923 (2653 - 360)$$
$$= 2476 \text{ kJ/kg}$$

Therefore,

$$V_2 = \sqrt{2(2951 - 2476) \times 10^3}$$
$$= 974.7 \text{ m/s}$$

Note – It is sufficiently accurate to round the value of h_2 to the nearest whole number, as the values in the table are quoted in this form.

5.5.3 Steady adiabatic flow through a throttle valve

A valve is represented by means of the schematic diagram shown in Figure 5.6.

In principle, a throttle valve is a device for reducing the pressure of the fluid flowing from state 1 to 2. This is particularly useful in the case of a refrigerator, where a throttle valve is a simple and low-cost device for allowing flow pressure to drop from the high-pressure side to the low-pressure side of the cycle. The drop in pressure can be achieved by simply putting a restriction in the flow, either in the form of an orifice or a capillary tube, as shown in Figure 5.7.

The throttle valve can be analysed using the steady flow energy equation, (5.5):

$$q + h_1 + \frac{V_1^2}{2} = h_2 + \frac{V_2^2}{2} + w$$

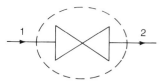

Figure 5.6 Schematic diagram of a valve.

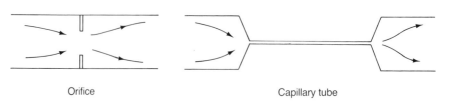

Orifice   Capillary tube

Figure 5.7 Types of restriction forming a throttle valve.

The following can be assumed.

1. There is no work done, so that $w = 0$.
2. The valve has negligible heat loss, or gain, so the flow through the valve is adiabatic and $q = 0$.
3. The mass flow rate across the valve is constant, so there is no change of velocity and $V_1 = V_2$.

Using these assumptions the steady flow energy equation for a throttle valve reduces to

$$h_1 = h_2 \tag{5.8}$$

Unlike either a diffuser or a nozzle, the flow through a throttle valve is irreversible. Flow from 1 to 2 results in a pressure drop. It is impossible to reverse the flow from 2 to 1 and increase the pressure. This would require an input of work that the throttle valve is incapable of doing. Since the process is irreversible, there is an increase in entropy, as illustrated earlier in Figure 3.10. This characteristic of a throttle valve allows it to be used to measure the state of a wet vapour.

Using straightforward measurements of pressure or temperature, the dryness fraction of a wet vapour cannot be found because the saturated temperature is dependent on the pressure. By expanding a sample of wet vapour through a throttle valve, the sample can be reduced to a sufficiently low pressure where the increase in entropy takes the vapour into the superheat region. In this state, the pressure and temperature can be used to define the enthalpy of the vapour. The following example illustrates the use of a throttle valve in this manner.

Example 5.8
Wet steam at a pressure of 1 MPa flows through a pipe. A sample of steam is expanded through a throttle valve down to a pressure of 100 kPa and leaves the valve with a temperature of 110°C. Use the outlet conditions of the throttle valve to estimate the dryness fraction of the steam in the pipe.

Conceptual model

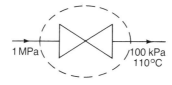

Process diagram – the flow is irreversible and the process can be represented on the T–s diagram. Analysis – from equation (5.8)

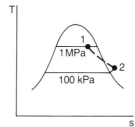

$$h_1 = h_2$$

The value of temperature T_2 is such that the value of h_2 must be found by interpolation between saturated vapour and a superheat temperature of 150°C. For steam at a pressure of 100 kPa, from Appendices A1 and A2,

$T = 99.6$°C, and $h_g = 2675$ kJ/kg for saturated vapour
$T = 150$°C, and $h = 2777$ kJ/kg for superheated steam

Interpolating between these values

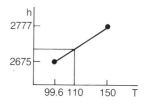

At 110°C

$$h_2 = 2675 + \frac{(110 - 99.6)}{(150 - 99.6)}(2777 - 2675)$$

$$= 2696\ \text{kJ/kg}$$

From equation (4.4)

$$h = h_f + x(h_g - h_f)$$

At state 1, taking properties from Appendix A1,

$$h_f = 763\ \text{kJ/kg}, \qquad h_g = 2778\ \text{kJ/kg}$$

Substituting,

$$2696 = 763 + x_1(2778 - 763)$$
$$x_1 = 0.96$$

Note – further calculation confirms that the entropy does increase for this irreversible expansion through the throttle valve as

$$s_1 = 6.408\ \text{kJ/kg K and } s_2 = 7.411\ \text{KJ/kg K}$$

5.5.4 Steady adiabatic flow with work

Compressors, turbines and pumps are all devices in which work crosses the boundary and in which the heat transfer to the surroundings can be considered to be negligible. The analysis of a pump has already been considered in connection with flow work, in section 5.2, so the present discussion will concentrate on the analysis of compressors and turbines.

Rotary compressors can be one of two types, centrifugal or axial, the name referring to the type of flow through the device. Figure 1.7 shows an engine with a typical centrifugal compressor while Figure 1.8 shows an engine with an axial flow compressor. Both operate on the same basic principle. The work input to the rotor causes an increase in the kinetic energy of the fluid. The fluid

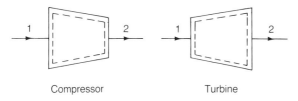

Compressor   Turbine

Figure 5.8 Schematic diagram of a compressor and turbine.

is then slowed down again in the static part of the compressor, where the extra kinetic energy is converted into a pressure increase.

Rotary turbines operate on the axial flow principle. The fluid enters a stationary series of nozzles in which the kinetic energy of the flow is increased to drive the rotor and produce a work output.

Figure 5.8 shows the schematic diagrams representing both a compressor and a turbine. It will be seen that one is the mirror image of the other, and in principle this is a useful way to view the two devices. A compressor is a device for increasing the pressure of a gas, or vapour, and it achieves this by work entering the compressor. A turbine is the reverse, it produces work by expanding a gas, or vapour, from a high pressure to a low pressure. Assuming that the processes are adiabatic and reversible, then in principle the turbine can be reversed to act as a compressor. Similarly, the compressor can be reversed to act as a turbine.

Both the compressor and the turbine can be analysed using the steady flow energy equation, (5.5):

$$q + h_1 + \frac{V_1^2}{2} = h_2 + \frac{V_2^2}{2} + w$$

The following can be assumed.

1. The device has negligible heat loss, so the flow is adiabatic and $q = 0$.
2. The change of velocity across the device is usually small, so that $V_1^2 \approx V_2^2$.

Using these assumptions, the steady flow energy equation reduces to the form:

$$w = h_1 - h_2 \tag{5.9}$$

Example 5.9
Air enters a compressor at a temperature of 20°C and is compressed through a pressure ratio of 10. Find the work done on each unit mass of air assuming the compression to be adiabatic and reversible. For air, take $\gamma = 1.4$ and $C_P = 1.005$ kJ/kg K.

Conceptual model

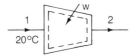

Process diagram

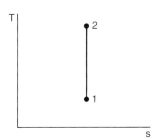

Analysis – the specific work done is given by equation (5.9):

$$w = h_1 - h_2$$

Since air can be assumed to be a perfect gas, from equation (4.13)

$$w = C_P (T_1 - T_2)$$

For an adiabatic process, the relationship between temperature and pressure is given by equation (4.25):

$$\frac{T_1}{T_2} = \left(\frac{P_1}{P_2}\right)^{(\gamma-1)/\gamma}$$

Re-arranging,

$$T_2 = T_1 \left(\frac{P_2}{P_1}\right)^{(\gamma-1)/\gamma}$$

$$= 293(10)^{0.4/1.4}$$

$$= 565.7 \text{ K}$$

Therefore,

$$w = 1.005 \times 10^3 (293 - 565.7)$$
$$= -274.1 \times 10^3 \text{ J/kg}$$
$$= -274.1 \text{ kJ/kg}$$

Note – the negative sign for the work means that it is a work **input** to the compressor.

Example 5.10
Steam enters a turbine at a pressure of 600 kPa and a temperature of 250°C. If it expands down to a pressure of 80 kPa calculate the work done for each kg of steam flowing through the turbine.

Conceptual model

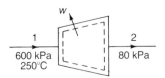

Process diagram

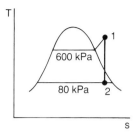

Analysis – the specific work done is given by equation (5.9):

$$w = h_1 - h_2$$

The steam enters in a superheated condition. From Appendix A2

$h_1 = 2958 \text{ kJ/kg}$
$s_1 = 7.182 \text{ kJ/kg K}$

At a pressure of 80 kPa, the saturated vapour has an entropy value of 7.434 kJ/kg K, from Appendix Al.
Since $s_2 = s_1 = 7.182$ kJ/kg K, this means that the steam at 2 is wet with a dryness fraction x_2.
From equation (4.5)

$$s_2 = s_f + x_2 (s_g - s_f)$$

At state 2, taking the properties from Appendix Al,

$s_f = 1.233 \text{ kJ/kg K}$ and $s_g = 7.434 \text{ kJ/kg K}$

Therefore,

$$7.182 = 1.233 + x_2(7.434 - 1.233)$$
$$x_2 = 0.959$$

From equation (4.4)

$$h = h_f + x(h_g - h_f)$$

At state 2, taking properties from Appendix Al,

$h_f = 392 \text{ kJ/kg}$ and $h_g = 2665 \text{ kJ/kg}$

Substituting,

$$h_2 = 392 + 0.959 (2665 - 392)$$
$$= 2571.8 \text{ kJ/kg}$$

Therefore,

$$w = 2958 - 2571.8$$
$$= 386.2 \text{ kJ/kg}$$

5.6 THE SFEE AS A RATE EQUATION

The steady flow energy equation as expressed in equation (5.5), relates to one unit mass of fluid flowing through the system. There is nothing to show whether 1 kg flows through the system in 1 second, or 1 minute. Therefore, equation (5.5) cannot be used to find the rate of change of energy within the system. However, it can be modified to provide a rate equation by introducing the mass flow rate into the steady flow energy equation.

Taking the steady flow energy equation, (5.5)

$$q + h_1 + \frac{V_1^2}{2} = h_2 + \frac{V_2^2}{2} + w$$

and multiplying throughout by the mass flow rate $\dot{m}$ gives

$$\dot{m}q + \dot{m}h_1 + \dot{m}\frac{V_1^2}{2} = \dot{m}h_2 + \dot{m}\frac{V_2^2}{2} + \dot{m}w$$

But the rate of heat transfer $\dot{m}q$ is expressed as Q, and the rate of work done $\dot{m}w$ is expressed as W. The rate equation can, therefore, be expressed as

$$Q + \dot{m}h_1 + \dot{m}\frac{V_1^2}{2} = \dot{m}h_2 + \dot{m}\frac{V_2^2}{2} + W \tag{5.10}$$

Since the terms in the steady flow energy equation are expressed in units of J/kg, the terms in the rate equation, (5.10), are expressed in units of

$$\frac{kg}{s} \times \frac{J}{kg} = \frac{J}{s} = W \text{ i.e. watts}$$

The rate equation (5.10) is particularly useful in analysing the rate of heat transfer in boilers, condensers or combustion chambers. Modifying equation (5.6) gives the rate of heat transfer as

$$Q = \dot{m}\,(h_2 - h_1) \tag{5.11}$$

Similarly, the rate of work done is the power and this can be evaluated for compressors or turbines using a modified form of equation (5.9):

$$W = \dot{m}\,(h_1 - h_2) \tag{5.12}$$

Example 5.11

A gas turbine operates with a mass flow rate of 10 kg/s of gas entering at 1250 K and a pressure of 1 MPa. If the gas leaves the turbine at a pressure of 100 kPa, calculate the power produced.

Assume the gas has the properties of air and behaves as a perfect gas with $\gamma = 1.4$, $C_P = 1.04$ kJ/kg K.

Conceptual model

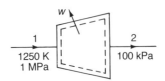

Process diagram – assuming the process to be adiabatic it can be represented on the T–s diagram.

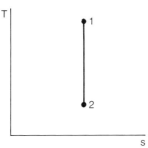

Analysis – the power is given by equation (5.12):

$$W = \dot{m} \, (h_1 - h_2)$$

Since the fluid is a perfect gas, from equation (4.13),

$$W = \dot{m} \, C_P \, (T_1 - T_2)$$

For an adiabatic process, the relationship between temperature and pressure is given by equation (4.25):

$$\frac{T_1}{T_2} = \left(\frac{P_1}{P_2}\right)^{(\gamma-1)/\gamma}$$

Re-arranging,

$$T_2 = T_1 \left(\frac{P_2}{P_1}\right)^{(\gamma-1)/\gamma}$$

$$= 1250 \left(\frac{100}{1000}\right)^{0.4/1.4}$$

$$= 647.4 \text{ K}$$

Substituting gives

$$W = \dot{m} \, C_P \, (T_1 - T_2)$$
$$= 10 \times 1.04 \, (1250 - 647.4)$$
$$= 6267 \text{ kW i.e. } 6.27 \text{ MW}$$

5.7   HEAT EXCHANGERS

A heat exchanger is a device in which heat transfer takes place between two fluids. A typical example is a car radiator. The jacket water from the engine is cooled in a heat exchanger using a flow of air. The air and jacket water are separated by means of metal pipe walls However, not all heat exchangers are made of metal. A range of materials including ceramics, plastics and even paper can be used. The important feature is that the two fluid flows are kept separate. Boilers and condensers are two important types of heat exchangers; these and other types are

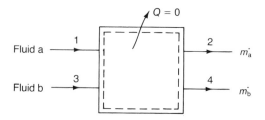

Figure 5.9 Schematic diagram of a heat exchanger.

discussed in more detail in Chapter 12. For the present, it is possible to portray a heat exchanger by means of the schematic diagram shown in Figure 5.9.

In Figure 5.9, there are two fluids 'a' and 'b' with two different flow rates, $\dot{m}_a$ and $\dot{m}_b$. It can be assumed that all the heat transfer takes place between the two fluid streams and there is no external heat transfer between the heat exchanger and the surroundings.

Because the two flow rates can be, and generally are, different it is necessary to analyse a heat exchanger using the form of the steady flow energy equation as a rate equation. It can be assumed that:

1. neither fluid flows do work, $w = 0$;
2. the velocities of each fluid flow, entering and leaving the heat exchanger, are equal, i.e. $V_1 = V_2$ and $V_3 = V_4$.

The rate equation for the heat exchanger can then be expressed in the form

$$Q + (\dot{m}_a h_1 + \dot{m}_b h_3) = (\dot{m}_a h_2 + \dot{m}_b h_4)$$

Since the rate of heat transfer to the surroundings is assumed to be zero, $Q = 0$:

$$\dot{m}_a(h_1 - h_2) = \dot{m}_b(h_4 - h_3) \tag{5.13}$$

Where the fluids are single phase, either liquids or gases, the change of enthalpy can be expressed in the form given in equation (4.13) as

$$(\dot{m}\,C_P)_a(T_1 - T_2) = (\dot{m}\,C_P)_b(T_4 - T_3) \tag{5.14}$$

Example 5.12
A car radiator cools 1kg/s of water from 98°C down to 80°C using air as the cooling fluid. If the air enters the radiator at 20°C with a velocity of 10 m/s and the frontal area of the radiator is 0.25 m², find the temperature of the air when it leaves. Take:

C_P for water $= 4.18$ kJ/kg K
C_P for air $= 1.005$ kJ/kg K
v for air $= 0.84$ m² /kg

Conceptual model

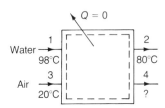

Analysis – from equation (5.l4) and using subscripts 'w' and 'a' for the water and air, respectively,

$$(\dot{m}\,C_P)_w\,(T_1 - T_2) = (\dot{m}\,C_P)_a\,(T_4 - T_3)$$

The mass flow rate of air is given by the conservation of mass, equation (5.l):

$$\dot{m} = \frac{VA}{v} = \frac{10 \times 0.25}{0.84} = 2.98\,\text{kg/s}$$

Substituting in the energy rate balance equation gives

$$1 \times 4.18\,(98 - 80) = 2.98 \times 1.005\,(T_4 - 20)$$
$$T_4 = 45.1°C$$

SUMMARY

In this chapter the application of the first law of thermodynamics to an open system has been discussed. The terms that have been introduced are:

steady flow
mass flow rate
flow work
steady flow energy equation
steady flow thermodynamic devices
boiler
condenser
combustion chamber
diffuser
nozzle
throttle valve
pump
compressor
turbine
SFEE as a rate equation
heat exchanger

Key equations that have been introduced.
Mass flow rate:

$$\dot{m} = VA/v \tag{5.1}$$

Flow work:

$$\text{flow work} = Pv \tag{5.2}$$

Steady flow energy equation:

$$q + h_1 + \frac{V_1^2}{2} = h_2 + \frac{V_2^2}{2} + w \tag{5.5}$$

Steady flow through a boiler, condenser or combustion chamber:

$$q = h_2 - h_1 \tag{5.6}$$

Velocity change in a diffuser or nozzle:

$$V_1^2 - V_2^2 = 2\,(h_2 - h_1) \tag{5.7}$$

Steady flow through a throttling valve:

$$h_1 = h_2 \tag{5.8}$$

Steady flow through a compressor or turbine:

$$w = h_1 - h_2 \tag{5.9}$$

SFEE as a rate equation:

$$Q + \dot{m}\,h_1 + \dot{m}\,\frac{V_1^2}{2} = \dot{m}\,h_2 + \dot{m}\,\frac{V_2^2}{2} + W \tag{5.10}$$

Rate equation for a boiler, condenser or combustion chamber:

$$Q = \dot{m}\,(h_2 - h_1) \tag{5.11}$$

Rate equation for a compressor or turbine:

$$W = \dot{m}\,(h_1 - h_2) \tag{5.12}$$

Rate equation for a heat exchanger:

$$\dot{m_a}(h_1 - h_2) = \dot{m_b}(h_4 - h_3) \tag{5.13}$$

PROBLEMS

1. Work output from a turbine is defined by the equation:

$$w = h_1 - h_2$$

State the assumptions used to derive this equation.

2. A gas enters a horizontal device with a temperature of 50°C, pressure of 150 kPa and a velocity of 100 m/s, through a pipe of 0.15 m diameter. The gas leaves with a temperature of 200°C, pressure of 200 kPa, through a pipe of 0.3 m diameter. Assume the gas constant to be 0.297 kJ/kg K. Calculate the velocity leaving the device.

3. Air enters a hair dryer at 20°C and leaves at 50°C through an outlet of 50 mm diameter. If the outlet velocity is 8 m/s calculate the rate of heat input to the dryer. Assume the air pressure to be 101 kPa and air to have $C_P = 1.005$ kJ/kg K and $R = 0.287$ kJ/kg K.

4. Steam at a pressure of 20 kPa and dryness fraction 0.9 enters a condenser. Assuming the steam leaves the condenser as saturated water, calculate the heat transferred during the process for each unit mass of steam.

5. Liquid hydrogen and liquid oxygen are burnt in a rocket engine to form highly superheated steam. The steam leaves the combustion chamber with negligible velocity and a temperature of 2000 K at a pressure of 800 kPa. If

it is expanded in a nozzle down to a pressure of 100 kPa, calculate the exit velocity. Assume the steam to behave as a perfect gas with $\gamma = 1.33$ and $C_P = 1.86$ kJ/kg K.

6. Refrigerant-134a enters a throttle valve as saturated liquid at a temperature of 30°C. If it expands to a saturated temperature of –5°C, find the condition at the outlet.

7. A turbine operates with steam entering at 1 MPa and 300°C. If the steam leaves the turbine at a pressure of 10 kPa, calculate the specific work output of the turbine.

8. Water enters a steam boiler at a temperature of 20°C and leaves at a pressure of 600 kPa with a dryness fraction of 0.85. If the steam flow is steady at 500 kg/hr and the enthalpy of the water entering is assumed to be 84 kJ/kg, find the rate of heat transfer in the boiler.

9. The condenser of a power station acts as a heat exchanger. Steam enters at 20 kPa with a dryness fraction of 0.8, and leaves as saturated liquid. It is cooled by water entering at 15°C. If the mass flow rate of steam is 200 kg/s, what is the required mass flow of cooling water if the water temperature cannot exceed 40°C? Take C_P for water as 4.18 kJ/kg K.

6 The second law of thermodynamics

6.1 AIMS

- To define a continuously operating heat engine.
- To use the Carnot analogy to define the second law of thermodynamics.
- To introduce the concept of thermal efficiency as applied to a heat engine.
- To use the Carnot conclusion to introduce the absolute temperature scale.
- To use the absolute temperature scale to introduce the concept of quality of energy.
- To show that the natural trend is for quality to decrease and entropy to increase.
- To introduce the reversed heat engine as a basis for a refrigerator.
- To introduce a continuously operating system operating on the Carnot cycle.
- To show that an irreversible process results in an increase in entropy.
- To introduce the adiabatic efficiency as applied to either a turbine or compressor.

6.2 HEAT ENGINES

The first law of thermodynamics is a statement of the principle of the conservation of energy. For both a closed system and an open system, energy can cross the boundary of the system in the form of heat and work. Both of these are forms of energy that are transferred from the system to the surroundings, or from the surroundings to the system. The result of this energy transfer brings about changes to the properties of the fluid within the system, but neither heat nor work are themselves properties.

It is possible to transfer heat to a system and, as a result, get a work output. A device especially designed to convert heat into work is called a 'heat engine'. The general development of heat engines was discussed in Chapter 1. The processes that take place within such heat engines can be analysed using the first law.

What is not clear from the first law is the relationship between work and heat within a heat engine. In other words, is it possible, at least in theory, to get a heat engine to produce the same work output as it receives heat input? In order to answer this question it is necessary to study the behaviour of heat engines.

Figure 6.1 shows a closed system consisting of a cylinder and frictionless piston. Assuming that the fluid inside the system to be water, heat input to the system will cause the water to go through a change of phase at constant pressure. First, the water will increase in temperature until it reaches saturation condition, at which point evaporation will take place. This will cause the piston to move,

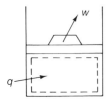

Figure 6.1 A simple heat engine.

doing work on the surroundings. The work output is a direct result of the heat input and the device can be considered to be a very simple form of heat engine.

As a heat engine, the work output will be less than the heat input, the difference being the change of internal energy within the system. However, as a heat engine, the system shown in Figure 6.1 is extremely limited. In fact, it is limited to just a single process because the work output must stop as soon as the piston reaches the end of its travel.

For this single process the work output is less than the heat input, but this is too limited an example to draw any general conclusions. It is possible to have a single process in which the work output is **greater** than the heat input, i.e. an adiabatic expansion. What is needed, is to consider a heat engine in which the operation is continuous.

The system shown in Figure 6.2 represents a simple steam engine of the type found in model shops. It consists of two flow devices, a boiler and an expander. The boiler is initially filled with water, which is converted to steam in order to drive the expander. The operation can be considered to be continuous whilst there is still steam in the boiler. Just as the simple engine, shown in Figure 6.1 is a closed system, the steam engine shown in Figure 6.2 can also be considered as a closed system, provided that the exhaust steam leaving the expander remains within the defined boundary.

It is fairly obvious that the work output of the steam engine must be less than the heat input to the boiler, because some of the energy will still be contained in the steam exhausting from the expander. This is demonstrated in the following example.

Figure 6.2 A simple steam engine.

Example 6.1
Compare the work output and heat input for a model boiler and steam engine. The boiler is filled with water at a saturation temperature of 99.6°C. This is evaporated and leaves the boiler as saturated vapour at a pressure of 400 kPa. The steam expands through the engine in a steady adiabatic flow process down to a pressure of 100 kPa.

Conceptual model – Figure 6.2.

Process diagram – define the state points as:

1. initial condition of the water;
2. steam leaving the boiler and entering the expander;
3. steam leaving the expander.

It is possible to represent the behaviour of the boiler and expander on the T–s diagram.

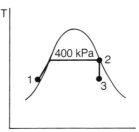

Analysis – the specific heat input to the boiler can be found from equation (5.6):

$$q = h_2 - h_1$$

From Appendix A1, $h_2 = 2739$ kJ/kg and $h_1 = 417$ kJ/kg.
Therefore

$$q = 2739 - 417 = 2322 \text{ kJ/kg}$$

Assuming that the adiabatic expansion from 2 to 3 is equivalent to adiabatic expansion through a turbine, the specific work output can be found from equation (5.9):

$$w = h_2 - h_3$$

From Appendix A1, h_2 is given above and $s_2 = 6.897$ kJ/kg K.
Now

$$s_3 = s_2 = 6.897 \text{ kJ/kg K}$$

So that

$$6.897 = 1.303 + x_3 \, (7.359 - 1.303)$$

and

$$x_3 = 0.924$$
$$h_3 = 417 + 0.924 \, (2675 - 417) = 2503 \text{ kJ/kg}$$

Therefore

$$w = 2739 - 2503 = 236 \text{ kJ/kg}$$

The ratio of the work output to the heat input is:

$$\frac{w}{q} = \frac{236}{2322} = 0.102$$

Note – The reason for the low work output compared to the heat input is the high enthalpy content, 2503 kJ/kg, of the steam leaving the engine. The work output as a ratio of the heat input is quantified by means of the thermal efficiency of the engine. In this example the efficiency is 10.2%.

6.3   THE SECOND LAW OF THERMODYNAMICS

Although the heat engine shown in Figure 6.2 has a work output that is less than the heat input, this does not prove that this is true for **all** heat engines. The generalized approach can be demonstrated using an analogy, first developed by Sadi Carnot in the early 1820s.

6.3.1   The Carnot analogy

Carnot lived at a time when heat was considered to be a colourless, weightless fluid called 'caloric'. He would have been familiar with water wheels and

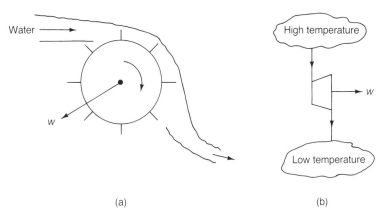

(a)   (b)

Figure 6.3 Analogy between a water wheel and a steam engine.

steam engines. He reasoned that both worked with a fluid, water in the case of water wheels and caloric in the case of steam engines. Similarly, a steam engine was considered to work with the caloric fluid falling from a high temperature source, a boiler, down to a low temperature receiver, a condenser, just as a water wheel works with water falling from a higher level to a lower one. This then was the basis of the Carnot analogy.

The Carnot analogy is illustrated in Figure 6.3. Figure 6.3(a) shows a water wheel in which water flows in at the top and out at the bottom. To ensure the conversation of mass, the flow of water out must equal the flow of water in, but in the process the water loses energy which is used to provide the work output from the wheel. However, the water flowing out must have energy, otherwise it would not flow away. Therefore, the water wheel cannot extract **all** the energy from the water flowing in.

Similarly, a heat engine as shown in Figure 6.3(b) operates between a high temperature reservoir and a low temperature reservoir. Heat energy flows into the engine from the high temperature reservoir. Just as there must be a flow of energy away from the water wheel, so there must be a flow of heat away from the engine to the low temperature reservoir. It follows that a heat engine must always have a work output that is, at best, equal to the difference between the heat flowing in and that flowing out. In other words, the work output will always be less than the heat input.

This conclusion is embodied in the second law of thermodynamics which can be stated as:

'It is impossible to make a continuously operating heat engine that converts all of the heat input into work output'.

This is known as the Kelvin–Planck statement of the second law of thermodynamics, and really says that the thermal efficiency of a continuously operating heat engine will always be less than 100%. In this connection it is important to emphasize that this applies to a continuously operating heat engine, not a single process heat engine as portrayed in Figure 6.1.

6.3.2 Thermal efficiency of a heat engine

From the statement of the second law, given above, it is possible to formally introduce the thermal efficiency of a continuously operating heat engine. Using the symbol η for the thermal efficiency, then

$$\eta = \frac{\text{work output}}{\text{heat input}}$$

In terms of the work output and heat input for one unit mass of the working fluid, the thermal efficiency can be expressed as

$$\eta = \frac{w_{\text{out}}}{q_{\text{in}}} \tag{6.1}$$

For a heat engine operating with a steady flow of the working fluid, it is possible to express the thermal efficiency in terms of the rate of work output, the power of the engine, as a proportion of the rate of heat input.

Assuming that the steady mass flow rate is $\dot{m}$, equation (6.1) can be modified to

$$\eta = \frac{\dot{m}\, w_{\text{out}}}{\dot{m}\, q_{\text{in}}}$$

and

$$\eta = \frac{W_{\text{out}}}{Q_{\text{in}}} \tag{6.2}$$

A schematic diagram of a continuously operating heat engine is shown in Figure 6.4. It is assumed to operate between two large thermal reservoirs at a high temperature T_{H}, and a low temperature T_{L}. If the heat input to the engine is q_{H}, for a unit mass of the working fluid, then the heat rejected from the engine to the low temperature reservoir is q_{L}. If the engine operates perfectly without friction, the work output must be the difference between the heat input and the heat output.

$$w_{\text{out}} = q_{\text{H}} - q_{\text{L}}$$

Substituting in equation (6.1)

$$\eta = \frac{q_{\text{H}} - q_{\text{L}}}{q_{\text{H}}}$$

and

$$\eta = 1 - \frac{q_{\text{L}}}{q_{\text{H}}} \tag{6.3}$$

Similarly, if the heat engine operates with a steady mass flow rate of the working fluid, the thermal efficiency can be expressed in terms of the rate of heat input and heat output:

$$\eta = 1 - \frac{Q_{\text{L}}}{Q_{\text{H}}} \tag{6.4}$$

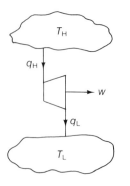

Figure 6.4 Schematic diagram of a heat engine.

Example 6.2

A heat engine has a thermal efficiency of 25% at a steady power output of 30 kW. Calculate the required fuel flow rate if the fuel has an energy content of 45 MJ/kg.

Conceptual model

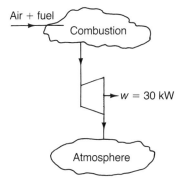

Process diagram – not defined for this problem.

Analysis – from equation (6.2)

$$\eta = \frac{W_{out}}{Q_{in}}$$

$$Q_{in} = \frac{30}{0.25} = 120 \, kW$$

To provide this rate of heat input the engine must use fuel at the rate of

$$\dot{m} = \frac{120 \times 10^3}{45 \times 10^0} = 0.00267 \, kg/s$$

Checking the units

$$\frac{W}{J/kg} = \frac{J/s}{J/kg} = \frac{kg}{s}$$

Note – a mass flow rate of 0.00267 kg/s is difficult to visualize. Assuming that the fuel is a liquid with a specific volume of 0.0012 m³/kg, the volume flow rate is

$$0.00267 \times 0.0012 = 3.2 \times 10^{-6} \, m^3/s$$
$$= 0.0115 \, m^3/hr \text{ i.e. } 11.5 \, l/hr$$

6.4   THE ABSOLUTE TEMPERATURE SCALE

The second outcome of Carnot's study of heat engines was the conclusion that the efficiency is independent of the type of working fluid used in an engine. A heat engine operating on steam between two thermal reservoirs will have exactly the same efficiency as one operating on air between the same temperatures. It is not the character of the working fluid but the magnitude of the temperatures of the thermal reservoirs that determine the efficiency.

Considering the heat engine shown in Figure 6.4, the engine operates between two thermal reservoirs at temperatures of T_H and T_L respectively. If the engine operates perfectly, without friction or losses, then the efficiency is given by equation (6.3):

$$\eta = 1 - \frac{q_L}{q_H}$$

Since the engine is assumed to be frictionless then it will be reversible and equation (6.3) can be taken as the expression for the efficiency of a reversible engine working between two thermal reservoirs at constant temperature.

As the heat input q_H increases with respect to the heat rejected q_L, the efficiency will improve. Therefore, the efficiency is a function of the ratio q_H/q_L:

$$\eta = f\left(\frac{q_H}{q_L}\right)$$

But Carnot's second principle states that the efficiency is a function of the temperatures of the reservoirs:

$$\eta = f\left(\frac{T_H}{T_L}\right)$$

From these relationships it can be shown that the ratio of the heat input to the heat rejected must be a function of the temperatures:

$$\frac{q_H}{q_L} = f\left(\frac{T_H}{T_L}\right)$$

There are several mathematical functions that would satisfy this equation. It was Lord Kelvin, then William Thomson, who proposed the relationship

$$\frac{q_H}{q_L} = \frac{T_H}{T_L} \tag{6.5}$$

This simple relationship defines the scale of absolute temperatures. On the Kelvin scale, the temperature ratio is independent of the working fluid. It simply depends on the ratio of the heat transfers between a reversible heat engine and the thermal reservoirs.

Taking the ratio of the temperatures given in equation (6.5) and substituting in equation (6.3) gives relationship for the efficiency:

$$\eta = 1 - \frac{T_L}{T_H} \tag{6.6}$$

This relationship gives the highest efficiency that can be achieved by a heat engine operating between two reservoir temperatures, T_H and T_L.

Example 6.3

Calculate the ideal thermal efficiency for the boiler and engine system defined in example 6.1.

Conceptual model and process diagram – as defined in example 6.1.

Analysis – the steam enters the expander at a saturation pressure of 400 kPa. From Appendix A1

$$T_H = 143.6°C = 416.6 \, \text{K}$$

The steam leaves at a pressure of 100 kPa. From Appendix A1

$$T_L = 99.6°C = 372.6 \, \text{K}$$

From equation (6.6)

$$\eta = 1 - \frac{T_L}{T_H} = 1 - \frac{372.6}{416.6} = 0.106 \quad \text{i.e. } 10.6\%$$

Note – This value of efficiency is slightly greater than that calculated in example 6.1. This is because the heat input, as defined in example 6.1, was not at one constant temperature but started at a temperature lower than the saturation temperature.

6.5   TEMPERATURE, EFFICIENCY AND ENTROPY

Defining the efficiency of a reversible heat engine in terms of the temperatures of the thermal reservoirs, allows the behaviour to be visualized. Temperature can be measured whereas heat input or rejection cannot be visualized in the same way.

Taking equation (6.6), it can be shown that efficiency improves when T_H is increased with relation to T_L. This can either be achieved by increasing T_H and/or reducing T_L. In the case of a steam engine this means increasing the boiler pressure and/or reducing the exhaust pressure leaving the engine. The latter can be achieved in practice by fitting a condenser to the outlet of the engine.

The development of heat engines, as outlined in Chapter 1, has been a quest for improved efficiency through increasing the ratio T_H/T_L. This is because efficiency has a direct bearing on the running costs of an engine. The efficiency is a measure of how much of the energy, originally contained in the fuel, is converted to useful work. An engine working on 30% will use two-thirds less fuel than one working on 10%. Not only does this mean that running costs are reduced because the more efficient engine uses less fuel, but that fuel reserves are conserved.

Therefore, the temperature ratio of T_H/T_L is not some obscure thermodynamic relationship but has a **direct bearing** on the development of heat engines used in present day society. In addition, it indicates that energy has 'quality' as well as quantity. The quality of energy is also sometimes referred to as its 'grade'.

A quantity of energy, say 1000 kJ, is the same whether it is held at 1000°C or at 200°C. However, the energy at 1000°C has a **much higher quality** than that at 200°C. To appreciate this, consider two reversible heat engines, one with $T_H = 1000$°C and the other with $T_H = 200$°C operating with the same low temperature reservoir at 20°C. The efficiency of the two engines can be compared as follows:

$$T_H = 1000°C, \ \eta = 1 - (293/1273) = 0.77$$
$$T_H = 200°C, \ \ \ \eta = 1 - (293/473) \ = 0.38$$

The energy at 1000°C has the potential to have 77% of the quantity converted to work, whereas the energy at 200°C only has a potential to have 38% converted. It can be concluded that the higher the temperature of an energy source, the higher its quality. High-quality energy has a greater potential to do useful work than low-quality energy.

In nature there is a tendency for energy to move from a higher quality state to a lower quality state. A cup of 'hot' tea, if left standing, will cool. Considering the cup and its surroundings, there will be no energy loss, but the tea in the cup will have gone from a higher temperature to a lower temperature – to a lower quality of energy.

There is no absolute measure of quality although temperature, as a measurable property, is a guide. The other property that provides an indication of the quality of energy is entropy. From the definition of entropy given in equation (3.7), it was shown that the heat transferred during an isothermal process was the area under the curve on a temperature–entropy diagram.

Consider the examples quoted above, in which two thermal reservoirs contain 1000 kJ of energy, one at 1000°C and the other at 200°C. They both have the potential to transfer 1000 kJ of energy in the form of heat. The heat transfer process at constant temperature for both thermal reservoirs is portrayed in Figure 6.5. In order to transfer the same quantity of heat, the change of entropy at the lower temperature 3–4, is considerably greater than that at the higher temperature 1–2. This means that the lower-temperature energy source starts with a greater value of entropy, at 3, than the high temperature source at 1. Since the natural trend is for energy sources to move from higher to lower quality, the same trend means that there is a tendency for entropy to increase.

The foregoing discussion has been based on the definition of efficiency that derives from the second law of thermodynamics and applies to a system and its surroundings. Although individual systems can undergo processes that result in a decrease in entropy, the total effect when the surroundings are taken into account is an increase in entropy.

Therefore, when considering a system and its surroundings in the light of the second law of thermodynamics, it can be stated that:

1. the natural trend is for the quality of energy to decrease;
2. the natural trend is for entropy to increase.

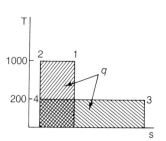

Figure 6.5 Comparison of heat transfer process.

6.6 REVERSIBLE HEAT ENGINE

A process in which entropy does not increase is during an adiabatic reversible process. During such a process the entropy remains constant. The idealized heat engine, defined in Figure 6.4, is assumed to operate perfectly without loss and without friction. In fact, equations (6.3) and (6.6) for the efficiency of a heat engine are based upon this assumption. Such a heat engine must operate on the basis of a reversible adiabatic process.

If a heat engine is reversible then its operation can be reversed. Instead of the heat input being used to create a work output, the engine will operate with a work input.

With a heat engine operating in a reversed direction, the direction of the heat input and heat rejection are also reversed, as shown in Figure 6.6. Heat is transferred from the low-temperature thermal reservoir and rejected to the high-temperature thermal reservoir. This is the basis of the refrigerator.

A refrigerator operates by absorbing heat from a low-temperature source and rejecting heat to a high-temperature source. This would be **impossible without the input of work**. Although the natural tendency is for heat to flow from a higher temperature to a lower temperature, a refrigerator does not infringe the second law of thermodynamics. The operation of a refrigerator gives rise to the Clausius statement of the second law:

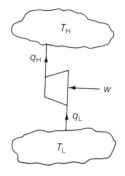

Figure 6.6 Reversed heat engine.

'It is impossible to make a continuously operating device whose only result is to transfer heat from a lower temperature reservoir to a higher temperature reservoir.'

Clearly, in the case of a refrigerator, the work input constitutes an effect other than the transfer of heat and ensures that the 'only result' in the Clausius statement is not infringed.

It is unusual to have two statements for the same law, the Kelvin–Planck statement given earlier and the Clausius statement. However, both are compatible as both stem from the behaviour of a heat engine operating in the way defined by Carnot.

The main difference is that whereas the Kelvin–Planck statement led to the definition of efficiency, no such performance criteria can be applied to a refrigerator. Since the sole purpose of a refrigerator is to transfer heat from a low-temperature source, the performance is determined by a coefficient of performance, defined as:

$$COP = \frac{\text{heat transferred}}{\text{work input}}$$

Applying the first law for a cycle to the refrigerator, the work input can be expressed in terms of the heat input and heat rejected as defined in Figure 6.6:

$$-w = q_L - q_H$$

Re-arranging gives the work input as

$$w = q_H - q_L$$

Since the sole purpose is to transfer heat from the low-temperature source, the coefficient of performance of a refrigerator can be defined as

$$COP = \frac{q_L}{q_H - q_L} \tag{6.7}$$

Using the relationship between the heat transfer and absolute temperature, given in equation (6.5), the coefficient of performance can be expressed in terms of the absolute temperatures of the thermal reservoirs:

$$COP = \frac{T_L}{T_H - T_L} \qquad (6.8)$$

If the refrigerator operates with a steady mass flow rate of the working fluid, the coefficient of performance can be expressed in terms of the rate of heat transfer from the low-temperature source and the power input:

$$COP = \frac{Q_L}{W} \qquad (6.9)$$

It should also be mentioned that a reversed heat engine can also operate as a 'heat pump'. This is a device for taking heat from a low-temperature source in order to transfer at a higher temperature for heating. The Festival Hall in London was originally heated using a heat pump, with the River Thames as the low-temperature source of the heat input.

Example 6.4

A refrigerator maintains a temperature of 2°C inside the cabinet while rejecting heat to the surrounding air at 25°C. If the rate of heat transfer is 200 W calculate the power input to the refrigerator.

Conceptual model:

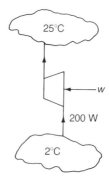

Analysis – assuming that the refrigerator operates perfectly, from equation (6.8)

$$COP = \frac{T_L}{T_H - T_L} = \frac{275}{298 - 275} = 11.95$$

Using this value of coefficient of performance, the power input can be found from equation (6.9):

$$W = \frac{Q_L}{COP} = \frac{200}{11.95} = 16.7\ W$$

Note – in practice the refrigerator would not have such a high coefficient of performance. Devices, such as refrigerator compressors, operate with considerable loss and a typical value of COP might be of the order of 3.

6.7 HEAT ENGINE OPERATING IN A CYCLE

The simple steam engine shown in Figure 6.2 was considered to operate continuously. However, this would only be true while there was steam in the boiler. As soon as the boiler ran dry, the engine would stop operating. To ensure that it operated in a truly continuous manner, it would be necessary to return the exhaust steam from the expander to the boiler, as illustrated in Figure 6.7.

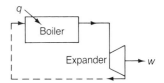

Figure 6.7 Modified boiler–steam engine system.

However, the pressure of the steam leaving the expander is lower than that in the boiler, so that it is impossible to construct a continuously operating system as shown in Figure 6.7. In order to get the exhaust steam back into the boiler, its pressure must be increased to that of the boiler. This can be achieved either by means of a compressor, in the case of vapour, or by condensing the steam into water and pumping back into the boiler.

A compressor operating between the exhaust pressure and the boiler pressure can be considered as a reversed expander, or reversed heat engine. Using this model, a continuously operating system can be envisaged as shown in Figure 6.8. This shows a heat engine, an expander, operating between a high-temperature and a low-temperature thermal reservoir. In parallel is a reversed heat engine, a compressor, to ensure that the working fluid returns from the low temperature to the high temperature thermal reservoir.

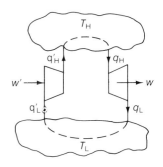

Figure 6.8 Continuously operating system.

The heat input to the expander is q_H and the heat rejection is q_L, resulting in work output w:

$$w = q_H - q_L$$

Similarly, the work input to the compressor is

$$w' = q'_H - q'_L$$

For the system to have a net work output then, the work output from the expander must be greater than the work input to the compressor:

$$(w)_{net} = w - w'$$
$$= (q_H - q_L) - (q'_H - q'_L)$$

Re-arranging gives

$$(w)_{net} = (q_H - q'_H) - (q_L - q'_L)$$
$$= (q_H)_{net} - (q_L)_{net}$$

Just as for the continuously operating heat engine on its own, the system shown in Figure 6.8 must have a net heat rejection to the low-temperature reservoir in order to achieve a net work output. For a steam engine this means that the steam leaving the expander must be partially condensed before entering the compressor. Similarly, for an engine working with a gas, the gas leaving the expander must be cooled before entering the compressor.

The thermal efficiency of the system shown in Figure 6.8 can be expressed in terms of the net work output compared to the net heat input:

$$\eta = \frac{(w)_{net}}{(q_H)_{net}}$$

It has been shown that the efficiency of a continuously operating heat engine, as shown in Figure 6.4, could be expressed in terms of the absolute temperatures of the thermal reservoirs. The system shown in Figure 6.8 also has an efficiency that can be expressed in the same way.

The system shown in Figure 6.8 can be represented as a cycle on a process diagram, in this case the T–s diagram shown in Figure 6.9. Since the operation of both a heat engine and a reversed heat engine, are independent of the working fluid, Figure 6.9 is true for any working fluid. The net heat input is an isothermal process at temperature T_H, whereas the net heat output is an isothermal process at temperature T_L.

It has been shown that the net work for the cycle is

$$(w)_{net} = (q_H)_{net} - (q_L)_{net}$$

so that the thermal efficiency for the cycle can be expressed as

$$\eta = \frac{(q_H)_{net} - (q_L)_{net}}{(q_H)_{net}}$$

For the two isothermal processes

$$(q_H)_{net} = T_H \times \Delta s$$

and

$$(q_L)_{net} = T_L \times \Delta s$$

Substituting in the relationship for the thermal efficiency above

$$\eta = \frac{T_H \times \Delta s - T_L \times \Delta s}{T_H \times \Delta s}$$

and the thermal efficiency of the cycle can be expressed in terms of the absolute temperature of the thermal reservoirs by cancelling out Δs:

$$\eta = \frac{T_H - T_L}{T_H} = 1 - \frac{T_L}{T_H}$$

This is the same relationship as for the continuously operating heat engine, equation (6.6). The cycle it is based on, Figure 6.9, is called the 'Carnot cycle'.

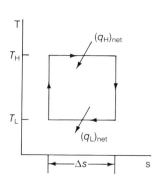

Figure 6.9 Process diagram for a continuously operating system.

Example 6.5

A power station operates on a Carnot cycle using steam between a boiler pressure of 1 MPa and a condenser pressure of 60 kPa. If the plant has a power output of 1 MW, calculate the heat rejection from the condenser.

Conceptual model

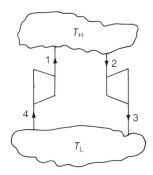

Process diagram

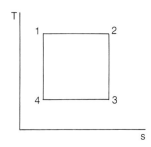

Analysis – from equation (6.6), for a Carnot cycle:

$$\eta = 1 - \frac{T_L}{T_H}$$

From Appendix A1:

at 1 MPa, $T_H = 179.9°C = 452.9$ K
at 60 kPa, $T_L = 86.0°C = 359$ K

Therefore

$$\eta = 1 - \frac{359}{452.9} = 0.207$$

The rate of heat input can be found from equation (6.2):

$$\eta = \frac{W_{out}}{Q_{in}}$$

$$0.207 = \frac{1 \times 10^6}{Q_{in}}$$

Therefore

$$Q_{in} = 4.82 \times 10 = 4.82 \text{ MW}$$

The rate of heat rejection is given by equation (6.4):

$$\eta = 1 - \frac{Q_L}{Q_H}$$

$$0.207 = \frac{Q_L}{4.82}$$

$$Q_L = 3.82 \text{ MW}$$

Note – the rate of heat rejection is also given by

$$Q_{in} - W_{out} = 4.82 - 1 = 3.82 \text{ MW}$$

6.8   REVERSIBILITY AND ENTROPY

A heat engine operating on a Carnot cycle between two thermal reservoirs at constant temperatures T_H and T_L is shown in Figure 6.10. Since the Carnot cycle is composed of **reversible** processes it is, therefore, **reversible**.

Assuming that the heat input to the cycle from the high temperature reservoir is q_H, and the heat rejection to the low temperature reservoir is q_L, the heat transfers can be related to the absolute temperatures of the thermal reservoirs through equation (6.5):

$$\frac{q_H}{q_L} = \frac{T_H}{T_L}$$

Re-arranging gives

$$\frac{q_H}{T_H} = \frac{q_L}{T_L}$$

It follows that for a heat engine operating on a reversible cycle

$$\frac{q_H}{T_H} - \frac{q_L}{T_L} = 0$$

This can be expressed as the sum of all the heat transfers divided by the respective absolute temperatures is zero for a reversible cycle:

$$\int_c \left(\frac{q}{T}\right)_{rev} = 0 \tag{6.10}$$

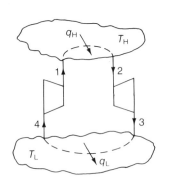

Figure 6.10 Heat engine operating on a Carnot cycle.

The question that arises from equation (6.10) is 'What happens to the sum of (q/T) in the case of a heat engine working on an irreversible cycle?' Consider an irreversible heat engine operating between the same thermal reservoirs as shown in Figure 6.10. Assuming that the irreversible heat engine receives the same input, q_H, then the heat rejected will be greater than that for a reversible heat engine. This can be shown by considering the work output:

$$(w)_{rev} = q_H - (q_L)_{rev}$$
$$(w)_{irrev} = q_H - (q_L)_{irrev}$$

Since one cause of irreversibilities is friction, the work output for an irreversible heat engine will be less than for a reversible heat engine

$$(w)_{rev} > (w)_{irrev}$$

This can only be true if

$$(q_L)_{rev} < (q_L)_{irrev}$$

It follows that for a heat engine operating on an irreversible cycle

$$\frac{q_H}{T_H} - \frac{(q_L)_{irrev}}{T_L} < 0$$

This can be expressed as

$$\int_c \left(\frac{q}{T}\right)_{irrev} < 0 \qquad (6.11)$$

Combining equations (6.10) and (6.11) gives a statement of what is called the 'Clausius inequality':

$$\int_c \left(\frac{q}{T}\right) \leq 0 \qquad (6.12)$$

The derivation of the statement of the Clausius inequality in equation (6.12) is important as it led to the definition of the property 'entropy'. However, since entropy has already been defined, in Chapter 3, it is not necessary to go over the same ground again. What is important is the use of entropy as **a criterion for irreversibility**.

Figure 6.11 shows two cycles for a heat engine, one reversible as given in Figure 6.11(a), the other irreversible as given in Figure 6.11(b). It is assumed that the part of the cycle that is irreversible is the expansion between 2 and 3i.

In the case of the reversible cycle in Figure 6.11(a), the change of entropy around the cycle is zero:

$$\frac{q_{12}}{T_{12}} - \frac{q_{43}}{T_{43}} = 0$$

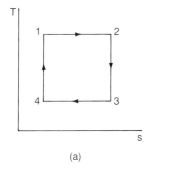
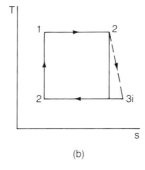

(a)   (b)

Figure 6.11 T–s diagrams of a reversible and an irreversible process.

that is

$$(\Delta s)_{12} - (\Delta s)_{43} = 0$$

However, in the case of the irreversible cycle in Figure 6.11(b), the change of entropy around the cycle is less than zero:

$$\frac{q_{12}}{T_{12}} - \frac{q_{43i}}{T_{43i}} < 0$$

that is

$$(\Delta s)_{12} - (\Delta s)_{43i} < 0$$

and

$$(\Delta s)_{43i} > (\Delta s)_{43}$$

This can only be true if the change of entropy between 2 and 3i is greater than the change of entropy between 2 and 3. It can be concluded that this irreversible process results in an increase in entropy. In fact, this is true for any irreversible process and the principle of an increase-in-entropy is the criterion for judging whether a process is reversible or irreversible.

It can be assumed that the isothermal heat input, or heat rejection, processes shown in Figure 6.11 are reversible. The main cause of irreversibility in a cycle is friction in the compression and expansion processes. Friction can occur in **both** these processes, unlike the cycle shown in Figure 6.11(b) where it was assumed that only the expansion process was irreversible.

To assess the effect of friction on the performance of a compressor or an expander, a component efficiency is introduced. In the case of steady flow the expander can be considered as a turbine. Both a compressor or a turbine operate with steady flow under adiabatic conditions. In other words, the heat transfer between the device and the surroundings is considered to be negligible. The component efficiency is, therefore, termed an 'adiabatic efficiency'.

Since the effect of friction is to reduce the work output of a turbine, the adiabatic efficiency defines the ratio of the actual work output compared to that if the expansion through the turbine was a reversible adiabatic process.

$$\eta_T = \frac{\text{actual work output}}{\text{reversible work output}} = \frac{w_a}{w_{rev}} \qquad (6.13)$$

Similarly, the effect of friction on a compressor is to increase the work input compared to a reversible adiabatic process.

$$\eta_c = \frac{\text{reversible work output}}{\text{actual work output}} = \frac{w_{rev}}{w_a} \qquad (6.14)$$

Example 6.6

Steam enters a turbine as saturated vapour at 800 kPa and leaves at a pressure of 60 kPa. If the turbine has an adiabatic efficiency of 90%, find the work output for a unit mass of steam and define whether the exhaust steam is wet or superheated.

Conceptual model

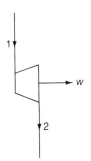

Process diagram

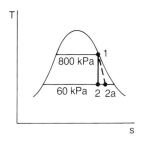

Analysis – with reversible adiabatic expansion, the work is given by equation (5.9):

$$w = h_1 - h_2$$

From Appendix A1

$h_1 = 2769$ kJ/kg and $s_1 = 6.663$ kJ/kg K.

At state 2

$s_2 = s_1 = 6.663$ kJ/kg K

From equation 4.5

$$s_2 = s_f + x_2(s_g - s_f)$$

$$6.663 = 1.145 + x_2 (7.531 - 1.145)$$

and

$$x_2 = 0.864$$

From equation (4.4)

$$\begin{aligned} h_2 &= h_f + x_2 (h_g - h_f) \\ &= 360 + 0.864 (2653 - 360) \\ &= 2341 \text{ kJ/kg} \end{aligned}$$

for the reversible expansion between 1 and 2

$$w = 2769 - 2341 = 428 \text{ kJ/kg}$$

However, the turbine operates with an adiabatic efficiency of 90%. From equation (6.13)

$$0.9 = \frac{w_a}{w_{rev}}$$

Therefore,

$$w_a = 0.9 \times w_{rev}$$
$$= 0.9 \times 428 = 385 \text{ kJ/kg}$$

The enthalpy at 2a is

$$h_{2a} = h_1 - w_a$$
$$= 2769 - 385 = 2384 \text{ kJ/kg}$$

Since this value of enthalpy is less than h_g at 60 kPa, it means that the exhaust steam is still wet!

Example 6.7

Air enters a rotary compressor at a temperature of 20°C. If the compressor operates with a pressure ratio of 10 and an adiabatic efficiency of 80%, calculate the temperature of the air at the outlet. Assume γ for air is 1.4.

Conceptual model

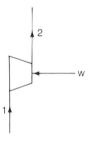

Process diagram

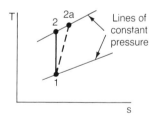

Analysis – With reversible adiabatic compression, the work is given by equation (5.9):

$$w = h_1 - h_2$$

Since air can be assumed to be a perfect gas, using equation (4.13) the work can be expressed as

$$w = C_P (T_1 - T_2)$$

From equation (4.25)

$$T_2 = T_1 \left(\frac{P_2}{P_1} \right)^{(\gamma - 1)\gamma}$$

$$= 293 \times (10)^{0.4/1.4}$$

$$= 565.7 \text{ K}$$

The compressor operates with an adiabatic efficiency of 80%. From equation (6.14)

$$0.8 = \frac{w_{rev}}{w_a} = \frac{C_P (T_1 - T_2)}{C_P (T_1 - T_{2a})}$$

The values of C_P cancel, so that the compressor efficiency can be expressed as

$$0.8 = \frac{(T_1 - T_2)}{(T_1 - T_{2a})} = \frac{293 - 565.7}{293 - T_{2a}}$$

that is

$$T_{2a} = 633.9 \text{ K}$$

SUMMARY

In this chapter the second law of thermodynamics has been defined through both the Kelvin–Planck statement,

'It is impossible to make a continuously operating heat engine that converts all of the heat input into work output'

and the Clausius statement,

'It is impossible to make a continuously operating device whose only result is to transfer heat from a lower temperature reservoir to a higher temperature reservoir'.

The key terms that have been introduced are:

heat engine
Carnot analogy
second law of thermodynamics
thermal efficiency
absolute temperature scale
quality of energy
entropy
reversible heat engine
refrigerator
coefficient of performance
Carnot cycle

Clausius inequality
adiabatic efficiency

Key equations that have been introduced.

Thermal efficiency:

$$\eta = \frac{w_{out}}{q_{in}} \tag{6.1}$$

$$\eta = \frac{W_{out}}{Q_{in}} \tag{6.2}$$

$$\eta = 1 - \frac{q_L}{q_H} \tag{6.3}$$

$$\eta = 1 - \frac{Q_L}{Q_H} \tag{6.4}$$

$$\eta = 1 - \frac{T_L}{T_H} \tag{6.6}$$

Coefficient of performance for a refrigerator:

$$COP = \frac{T_L}{T_H - T_L} \tag{6.8}$$

$$COP = \frac{Q_L}{W} \tag{6.9}$$

Adiabatic efficiency:

of a turbine, $$\eta_T = \frac{w_a}{w_{rev}} \tag{6.13}$$

of a compressor, $$\eta_C = \frac{w_{rev}}{w_a} \tag{6.14}$$

PROBLEMS

1. An inventor claims to have invented a steam engine operating between a boiler pressure of 1 MPa and a condenser pressure of 40 kPa, having a thermal efficiency of 25%. Is this possible?

2. An industrial gas turbine engine develops a power output of 8 MW at a thermal efficiency of 28%. Determine the required mass flow rate of the fuel if the fuel has an energy content of 46 000 kJ/kg.

3. A heat engine works between the temperature limits of 600°C and 100°C. When developing 40 kW it consumes 12 kg/hr of fuel. If the fuel has an

energy content of 40 000 kJ/kg, calculate the ratio of the actual thermal efficiency with relation to the Carnot efficiency.

4. A car engine has an output of 30 kW when working at a thermal efficiency of 30%. If half the heat rejected is passed to the cooling water, find the water flow rate to limit the cooling water temperature rise to 15°C. Take C_P for water as 4.2 kJ/kg K.

5. A refrigerator operates with a temperature of 0°C inside the cabinet when the ambient temperature of the surrounding air is 30°C. If the motor delivers a power input of 50 W to the refrigerator, calculate the rate of heat transfer from the cabinet.

6. A solar powered heat engine operates on a Carnot cycle with R-134a as the working fluid. If the engine has a power output of 500 kW and operates between a solar collector pressure of 1012.2 kPa and a condenser pressure of 487.3 kPa, calculate the heat rejection to the atmosphere.

7. A coal-fired power station operates on a Carnot cycle with steam at a boiler pressure of 2 MPa and a condenser pressure of 40 kPa. If the station has a power output of 600 MW, calculate the required coal consumption in tonnes/hr if the coal has an energy content of 32 000 kJ/kg. Take one tonne $= 10^3$ kg.

8. A model steam engine operates with isothermal heat input to the boiler at a pressure of 400 kPa and expansion down to an exhaust pressure of 100 kPa. Assuming that the boiler operates between saturated water and a dryness fraction of 0.9, and the adiabatic efficiency of the expansion process is 70%, calculate the thermal efficiency of the engine.

9. The compressor of an aircraft engine takes in air at 10°C, operates with a pressure ratio of 25 and has an adiabatic efficiency of 84%. If the mass flow rate of air through the compressor is 20 kg/s, find the power input required. Assume that air has $C_P = 1.03$ kJ/kg K and $\gamma = 1.4$.

7 Vapour cycles

7.2 STEAM POWER CYCLES

Steam has been used as a working fluid for heat engines, for some three centuries. Starting with the Newcomen engine, steam was condensed in a cylinder to form a partial vacuum to allow atmospheric pressure to drive a piston. High-pressure steam, i.e. above atmospheric pressure, was first used to drive reciprocating engines and then rotating turbines. During this period the application of steam engines to factories, railways and power stations has brought about a social revolution. However, these changes have been achieved at a cost, that cost being the use of fossil fuels.

It is not proposed to take another look at the history of steam engines. That has already been briefly outlined in section 1.4. At this point it is intended to look at steam engines in the context of the findings from the last chapter. In particular, to use thermal efficiency as an important performance criterion for judging the effectiveness, or otherwise, of a steam engine. As mentioned in section 6.5, thermal efficiency is a measure of how effectively the energy contained in a fuel is converted to useful work.

The early Newcomen engine operated with a thermal efficiency of well below 1%. A modern power station operates on a steam cycle with a thermal efficiency of, typically, 35%. The increase in the thermal efficiency over the

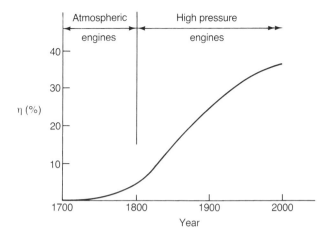

Figure 7.1 Improvement in the thermal efficiency of steam engines.

period is shown in Figure 7.1. In three centuries the thermal efficiency has broadly improved by a factor of fifty. A power station working on the same low efficiency as the Newcomen engine would burn about fifty times the amount of coal that is currently used. This means that not only would costs increase by a factor of fifty, but also that coal reserves would be squandered at a more rapid rate. Improving thermal efficiency, therefore, not only reduces running costs but also helps conserve energy reserves.

The idealized heat engine, as portrayed in the last chapter, was considered to operate between two thermal reservoirs, one at high temperature and the other at low temperature. In the case of a steam engine the high-temperature reservoir is the boiler. The low-temperature reservoir is achieved by exhausting steam to either the atmosphere, as in the case of a steam locomotive, or a condenser. Using the ideal heat engine efficiency as given in equation (6.6):

$$\eta = 1 - \frac{T_L}{T_H}$$

it can be shown that the thermal efficiency improves by increasing the temperature T_H or reducing the temperature T_L.

Operating with lower condenser pressures improves the thermal efficiency because the saturation temperature decreases with the pressure. Similarly, increasing boiler pressure improves the thermal efficiency. The trend in developing steam engines has been to do just that; improve the thermal efficiency through increasing boiler pressures while maintaining low condenser pressures. Whereas steam engines had boiler pressures of about 400 kPa in the early nineteenth century, some power stations now operate on boiler pressures as high as 20 MPa.

The increase in the **potential** ideal thermal efficiency brought about by this increase in boiler pressure is quite significant. Comparing these two boiler pressures for a constant condenser pressure of, say, 20 kPa, i.e. $T = 60.1°C$:

Pressure	T_H	$1 - T_L/T_H$
400 kPa	143.6°C	0.20
20 MPa	365.8°C	0.48

In practice, the actual efficiencies are less than the ideal thermal efficiency due to the irreversible behaviour of real components, such as a reciprocating engine or rotating turbine. Nevertheless, the ideal thermal efficiency is a useful concept in defining those thermodynamic changes that are necessary to achieve improvements in the actual performance of steam cycles.

Thermal efficiency is an important criterion for judging the performance of actual steam plant, as it is a measure of how effectively the fuel is used. It, therefore, provides an inverse indication of the running costs incurred. The higher the efficiency, the lower the fuel consumption and the lower the running costs. However, this is just one performance criterion, the other being the specific work output.

The specific work output is the measure of the net work from a cycle for a unit mass of working fluid, w_{net}. As such, it provides an inverse measure of the required mass flow rate to achieve a given power output. The size of the components, comprising the steam plant, will depend on the magnitude of the mass flow rate. The size will govern the capital cost involved and so the specific work output is an important criterion as it gives an inverse indication of the capital cost of the plant.

7.3   THE CARNOT CYCLE

The Carnot cycle, discussed in the last chapter, is the most efficient cycle operating between two constant temperature thermal reservoirs. Since it gives promise of the highest thermal efficiency, it is sensible to study the Carnot steam cycle before looking at any alternative cycle. In the case of a Carnot steam cycle the heat addition and heat rejection at constant temperature, are achieved in a boiler and condenser respectively.

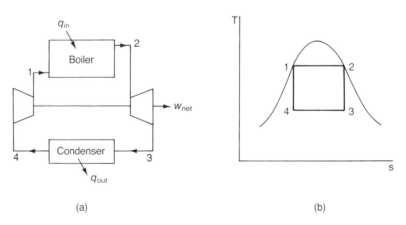

(a)   (b)

Figure 7.2 The Carnot steam cycle.

The system required to achieve a Carnot cycle operating on steam, or any other suitable two-phase fluid, is shown in Figure 7.2(a). The system shown in Figure 7.2(a) is a closed system. The boundary has been left out for clarity. However, the boundary is still there with heat and work crossing between the components and the surroundings. The resulting cycle is shown on the T–s diagram in Figure 7.2(b). It consists of four processes:

1–2, isothermal heat addition in a boiler;
2–3, reversible adiabatic expansion;
3–4, isothermal heat rejection in a condenser;
4–1, reversible adiabatic compression.

The diagram of the cycle in Figure 7.2(b) shows the heat addition to be at constant temperature from the saturated liquid to the saturated vapour state. This process ensures the maximum heat input at a given boiler pressure and, therefore, the maximum work output from the cycle.

Thermal efficiency for the cycle has already been defined in the last chapter by equation (6.6)

$$\eta = 1 - \frac{T_L}{T_H}$$

where the high temperature

$$T_H = T_1 = T_2$$

and the low temperature

$$T_L = T_3 = T_4$$

Since the efficiency is a measure of the net work output compared to the heat input, the specific work output for the cycle can be defined as

$$w_{net} = \eta \, (q_{in})$$

But the heat input in the boiler can be expressed in terms of the enthalpy change from state 1 to state 2. From equation (5.6)

$$q_{in} = h_2 - h_1$$

and the specific work output can be expressed as

$$w_{net} = \eta \, (h_2 - h_1) \tag{7.1}$$

Example 7.1
A steam plant operates on a Carnot cycle between a boiler pressure of 1 MPa and a condenser pressure of 20 kPa. Find the thermal efficiency and specific work output for the plant.

Conceptual model – Figure 7.2(a).

Process diagram – Figure 7.2(b).

Analysis – at a boiler pressure of 1 MPa, from Appendix A1

$$T_H = 179.9°C = 452.9 \text{ K}$$

Similarly, at a condenser pressure of 20 kPa,

$$T_L = 60.1°C = 333.1 \text{ K}$$

From equation (6.6)

$$\eta = 1 - \frac{T_L}{T_H} = 1 - \frac{333.1}{452.9} = 0.265$$

From equation (7.1)

$$w_{net} = \eta (h_2 - h_1)$$

From Appendix A1

$$h_2 = 2778 \text{ kJ/kg}$$
$$h_1 = 763 \text{ kJ/kg}$$

Therefore

$$w_{net} = 0.265 (2778 - 763) = 533 \text{ kJ/kg}$$

7.4   THE RANKINE CYCLE

7.4.1   Limitations of the Carnot cycle

In practice, it is impossible to achieve the Carnot cycle as portrayed in Figure 7.2(b). The process from state 4 to state 1 consists of a reversible adiabatic compression from a wet steam condition at 4 to saturated water at 1. In the first place it is impractical to design a compressor that would achieve this change of phase; in the second place it is extremely difficult to control the condensation process so that the condition of the wet steam ends precisely at state 4.

Rather than modify the Carnot cycle to try and overcome these limitations, a more practical approach is to ensure that the condensation process continues until all the exhaust steam is changed to water. The water can then be pumped back into the boiler to complete the cycle. This is the basis of the Rankine cycle.

7.4.2   The Rankine steam cycle

The Rankine cycle is the basis for all engines or plant operating on steam.

The plant required to achieve a Rankine cycle is shown in Figure 7.3(a). It consists of a boiler, expander, condenser and pump. For most practical applications the expander can be considered to be a rotating turbine. The resulting cycle is shown on the T–s diagram in Figure 7.3(b). It consists of four processes:

1–2, heat addition in a boiler at constant pressure;
2–3, reversible adiabatic expansion from boiler pressure down to condenser pressure;
3–4, isothermal heat rejection in a condenser to saturated water state;
4–1, reversible adiabatic pumping from condenser pressure back to boiler pressure.

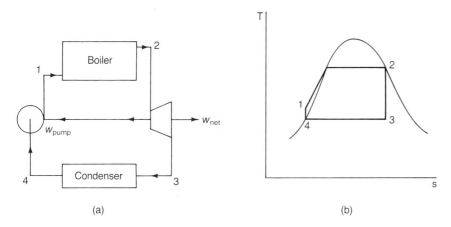

Figure 7.3 The Rankine steam cycle.

The thermal efficiency for the Rankine cycle can be expressed in terms of the net work output compared to the heat input:

$$\eta = \frac{w_{net}}{q_{in}} = \frac{w_{23} - w_{41}}{q_{12}} \qquad (7.2)$$

Since all the components of the steam plant in Figure 7.3(b) involve steady flow, the work of the turbine and pump can be found by applying the steady flow energy equation. From equation (5.9), the work output from the turbine is

$$w_{23} = h_2 - h_3$$

From equation (5.2) for the flow work, the work input to the pump is

$$w_{41} = v(P_4 - P_1)$$
$$= -v(P_1 - P_4)$$

From equation (5.6), the heat input in the boiler is

$$q_{12} = h_2 - h_1$$

Combining these gives the following relationship for the thermal efficiency:

$$\eta = \frac{h_2 - h_3 - v(P_1 - P_4)}{h_2 - h_1} \qquad (7.3)$$

Example 7.2
A steam plant operates on a Rankine cycle between a boiler pressure of 1 MPa and a condenser pressure of 20 kPa. Find the thermal efficiency and specific work output for the plant, assuming the steam leaves the boiler as a saturated vapour.

Conceptual model – Figure 7.3(a).

Process diagram – Figure 7.3(b).

Analysis – at state 2, from Appendix A1,

$$h_2 = 2778 \text{ kJ/kg}, \quad s_2 = 6.586 \text{ kJ/kg K}$$

At state 3,

$$s_3 = s_2 = 6.586 \text{ kJ/kg K}$$

Therefore

$$6.586 = s_f + x_3 (s_g - s_f)$$

Taking values of specific entropy from Appendix A1,

$$6.586 = 0.832 + x_3 (7.907 - 0.832)$$
$$x_3 = 0.813$$

Taking values of specific enthalpy from Appendix A1,

$$h_3 = 251 + 0.813 (2609 - 251)$$
$$= 2168 \text{ kJ/kg}$$

At state 4,

$$h_4 = 251 \text{ kJ/kg}$$

At state 1, it is not possible to find the enthalpy value directly from the table. It has to be calculated assuming that

$$w_{41} = h_4 - h_1$$

Therefore

$$h_1 = h_4 - w_{41}$$
$$= h_4 - (-v(P_1 - P_4))$$
$$= 251 + 10^{-3} (1000 - 20)$$
$$= 252 \text{ kJ/kg}$$

Using these values in equation (7.3),

$$\eta = \frac{h_2 - h_3 - v(P_1 - P_4)}{h_2 - h_1}$$

$$= \frac{2778 - 2168 - 10(1000 - 20)}{2778 - 252}$$

$$= 0.241$$

The specific work output is given by:

$$w_{net} = w_{23} + w_{41}$$
$$= h_2 - h_3 - v(P_1 - P_4)$$
$$= 2778 - 2168 - 10^{-3} (1000 - 20)$$
$$= 609 \text{ kJ/kg}$$

Note – Before going on to look at the Rankine cycle in greater detail, it is instructive to see the part 'entropy' has played in this example. Entropy was introduced in Chapter 3 as a useful thermodynamic property. It has already been widely used in process diagrams. This example illustrates an essential role for entropy. It would be impossible to analyse a steam cycle of this nature without this property. The assumption that $s_2 = s_3$ and the use of that assumption to find the condition of the steam at state 3 is central to the analysis.

7.4.3 Comparison of the Carnot and Rankine cycles

A comparison between the results of examples 7.1 and 7.2 shows that, between the same boiler and condenser pressure, the thermal efficiency of the Carnot cycle is higher than the Rankine cycle, i.e. 26.5% compared to 24.1%. This is what would be expected because the Carnot cycle is the **most efficient** that can be achieved between two given constant temperature thermal reservoirs. In this case, the reduced thermal efficiency for the Rankine cycle is due to the fact that heat input is not at constant temperature. The feed water inlet at state 1 is well below the saturation temperature in the boiler.

However, the lower thermal efficiency of the Rankine cycle is offset by a much greater specific work output, 609 kJ/kg compared to 533 kJ/kg for the Carnot cycle. The reason for this is that the work input to the pump in the Rankine cycle is very much smaller than the work input to the compressor in the Carnot cycle, i.e. 1 kJ/kg as compared to 77 kJ/kg.

This high work input to the compressor is even more significant if the adiabatic efficiency of the components is taken into account. The analysis of both the Carnot cycle, in example 7.1, and the Rankine cycle, in example 7.2, has been based upon the assumption that the turbine, compressor and pump operate perfectly. In practice, they would have adiabatic efficiencies of less than 1.

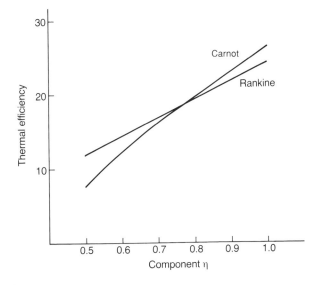

Figure 7.4 Comparison of thermal efficiency with component efficiency.

If the adiabatic efficiencies of the components are taken into account when analysing the basic Carnot and Rankine cycles, the effect on the thermal efficiency is shown in Figure 7.4. Assuming a constant value of adiabatic efficiency for each component, it will be seen that for a component efficiency of less than 0.8, the thermal efficiency of the modified Carnot cycle actually falls below that for the Rankine cycle.

Therefore, the Rankine cycle is a far more realistic cycle than the Carnot cycle as a basis for a steam engine, or plant. Not only does the thermal efficiency not suffer as great a drop when incorporating real components, turbines and pumps, but the specific work output is **greater** than for the Carnot cycle.

7.5 DEVELOPMENT OF THE RANKINE CYCLE

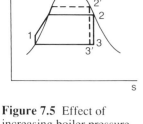

Figure 7.5 Effect of increasing boiler pressure.

The thermal efficiency of a steam plant working on the Rankine cycle can be improved by increasing the boiler pressure. The effect of such an increase, for a given condenser pressure, is shown in Figure 7.5.

Taking a Rankine cycle as defined by states 1–2–3–4, an increased boiler pressure results in the cycle defined by states 1–2′–3′–4. The latter cycle would have both a higher thermal efficiency and a greater specific work output compared to the former. However, there is one other factor that has to be considered. The exhaust steam at the end of the expansion has a lower dryness fraction at state 3′, than the original cycle at state 3.

This means that there is more saturated water in the steam at state 3′ than at state 3. Assuming that the expansion takes place within a turbine, the steam flows through the nozzles and rotating blades at a high velocity, typically up to 500 m/s. Any water droplets in the steam at this velocity will cause erosion and damage to the blades. The wetter the steam the greater the damage.

To minimize the damage to the blades it is essential to move the state of the exhaust steam towards the saturated vapour line, i.e. to the right on the T–s diagram. This can be achieved by moving the state of the steam into the turbine to the right, which means the steam leaving the boiler in a superheated state, as illustrated in Figure 7.6.

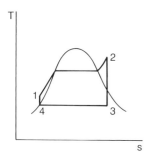

Figure 7.6 The Rankine cycle with superheat.

Clearly the higher the superheat temperature at state 2, shown in Figure 7.6, the drier the steam is going to be at the outlet of the turbine. However, there is a limit to which the superheat temperature can be raised. The materials used in the construction of steam turbines limit the maximum temperature to about 600°C. Fossil fuel power stations currently operate with superheat temperatures of up to 565°C.

7.5.1 The Rankine cycle with superheat

The analysis of the Rankine cycle with superheat can be carried out in exactly the same way as for the Rankine cycle considered in example 7.2. The work output from the turbine is again given by equation (5.9):

$$w_{23} = h_2 - h_3$$

Similarly, the input to the boiler is given by equation (5.6):

$$q_{12} = h_2 - h_1$$

From example 7.2, it was shown that the work input to the pump was a very small proportion of the work output from the turbine. In that example about 0.16%. In practice, this is typical of steam power plant, the pump work being generally less than 1% of that from the turbine. It can, therefore, be safely ignored in the analysis of the Rankine cycle. This leads to the assumption that the work output from the turbine is the net work output from the cycle. Also, that the increase in enthalpy from state 4 to state 1 is negligible:

$$h_1 \approx h_4$$

Incorporating these assumptions, the thermal efficiency of the Rankine cycle can be expressed as

$$\eta = \frac{w_{net}}{q_{in}} \approx \frac{h_2 - h_3}{h_2 - h_4} \tag{7.4}$$

Example 7.3

A steam power plant has steam leaving the boiler at 1 MPa and 300°C. If the plant operates with a condenser pressure of 20 kPa, calculate the thermal efficiency and specific work output for the cycle. Ignore the feed pump work in the analysis.

Conceptual model – Figure 7.3(a).

Process diagram – Figure 7.6.

Analysis – at state 2, from Appendix A2,

$$h_2 = 3052 \text{ kJ/kg and } s_2 = 7.124 \text{ kJ/kg K}$$

At state 3,

$$s_3 = s_2 = 7.124 \text{ kJ/kg K}$$

Therefore

$$7.124 = s_f + x_3 (s_g - s_f)$$

Taking values of specific entropy from Appendix A1,

$$7.124 = 0.832 + x_3 (7.907 - 0.832)$$
$$x_3 = 0.89$$

Taking values of specific enthalpy from Appendix A1,

$$h_3 = 251 + 0.89 (2609 - 251)$$
$$= 2350 \text{ kJ/kg}$$

At state 1,

$$h_1 \approx h_4 = 251 \text{ kJ/kg}$$

Using these values in equation (7.4):

$$\eta = \frac{h_2 - h_3}{h_2 - h_4}$$

$$= \frac{3052 - 2350}{3052 - 251} = 0.251$$

The specific work is given by equation (5.9):

$$w_{net} = h_2 - h_3$$
$$= 3052 - 2350 = 702 \text{ kJ/kg}$$

Note – compared to the Rankine cycle analysed in example 7.2, the effect of superheating is to improve the thermal efficiency, from 24.1% to 25.1%, and increase the specific work output, from 609 to 702 kJ/kg.

7.5.3 Rankine cycle with two-stage expansion

Any increase in the temperature at which the heat is transferred to the cycle results in an improvement to the thermal efficiency. This is why the Rankine cycle with superheat has an improved thermal efficiency compared to the basic cycle, see example 7.3 above.

However, although superheating causes an increase in the dryness fraction of the steam at the outlet of the turbine from 0.813, in example 7.2, to 0.89, in example 7.3, the resulting value is still well below the saturated vapour condition. Increasing boiler pressure would reduce the dryness fraction still further.

Therefore, superheating on its own is not a cure for increasing the dryness fraction of the steam through the turbine. This can be achieved by splitting the expansion of the steam into several stages with the steam being reheated in the boiler between each stage. A modern power station utilizes large steam turbine–generator sets with three stages of expansion through a high-pressure turbine, an intermediate-pressure turbine and low-pressure turbine. For the present discussion it is sufficient to briefly describe a cycle with two-stage expansion, as shown in the process diagram given in Figure 7.7.

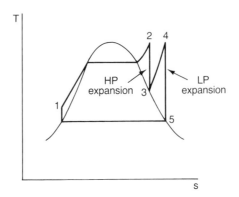

Figure 7.7 Steam cycle with two-stage expansion.

Expansion starts in the high-pressure (HP) turbine 2–3, until the steam reaches near saturated vapour condition. The steam at this lower pressure is then reheated in the boiler, generally to the same temperature as the super-heated steam at state 2, before being finally expanded in a low-pressure (LP) turbine 4–5. The reheating of the steam between the HP and the LP stages, ensures that the dryness fraction of the exhaust steam at 5 is improved. With this emphasis on the importance of reheating, the cycle shown in Figure 7.7 is also referred to as the 'Rankine cycle with reheat'.

The analysis of the cycle shown in Figure 7.7 is outside the scope of this book but can be found in more advanced texts; for example Look and Sauer (1988), and Eastop and McConkey (1986).

7.6   REVERSED CARNOT CYCLE

The Carnot cycle, as illustrated in Figure 7.1, consists of four processes. Two of these are isothermal heat transfer processes, which are reversible. The other two processes are reversible adiabatic expansion and compression. Because all the processes are reversible, the whole cycle can be reversed with associated changes in the direction of the heat and work transfer. Such a reversed Carnot cycle is the basis of a refrigerator.

A reversed Carnot cycle is shown in Figure 7.8. The devices making up the system required to achieve a reversed Carnot cycle are shown in Figure 7.8(a). Heat input to the low-temperature side of the cycle is achieved through a boiler. However, to distinguish between a boiler used in a steam power cycle and that used in a refrigerator, it is now called an 'evaporator'. Since the system oper-ates in a reverse direction to that shown in Figure 7.1, the flow through the system is anti-clockwise. Similarly, the resulting cycle shown in Figure 7.8(b), operates in an anti-clockwise direction. It consists of four processes:

1–2, reversible adiabatic compression;
2–3, isothermal heat rejection to the surroundings;
3–4, reversible adiabatic expansion;
4–1, isothermal heat input from the refrigeration space.

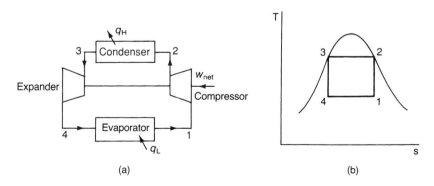

(a)   (b)

Figure 7.8 Reversed Carnot cycle.

The diagram of the cycle in Figure 7.8(b) shows the heat rejection to be at constant temperature from the saturated vapour to the saturated liquid state. This ensures the maximum heat input to the evaporator between states 4 and 1.

The performance of a refrigeration cycle is no longer judged in terms of the thermal efficiency but, as explained in section 6.6, in terms of a coefficient of performance. Since the purpose of a refrigeration cycle is to transfer heat from a low temperature store, the coefficient of performance can be defined as

$$COP = \frac{\text{heat input to the evaporator}}{\text{net work input}}$$

$$= \frac{q_{in}}{w_{net}}$$

Applying the first law of thermodynamics to the reversed Carnot cycle, the net work input can be expressed in terms of the heat input and heat rejection, as defined in Figure 7.8(a):

$$-w_{net} = q_L - q_H$$

Re-arranging gives the work input as

$$w_{net} = q_H - q_L$$

The coefficient of performance for a reversed Carnot cycle can be expressed as

$$COP = \frac{q_L}{q_H - q_L} \qquad (7.5)$$

Since the ratio of the heat input and heat rejection is equal to the absolute temperatures of the evaporation and condensation processes, the coefficient of performance can be defined using equation (6.8):

$$COP = \frac{T_L}{T_H - T_L}$$

where the low temperature

$$T_L = T_4 = T_1$$

and the high temperature

$$T_H = T_2 = T_3$$

Example 7.4
A refrigerator operates on a reversed Carnot cycle with Refrigerant-12 as the working fluid between an evaporator temperature of −5°C and a condenser temperature of 35°C. Calculate the coefficient of performance and the heat transferred to the evaporator.

Conceptual model – Figure 7.8(a).

Process diagram – Figure 7.8(b).

Analysis as defined in the question,

$$T_H = 35°C = 308 \text{ K}$$
$$T_L = -5°C = 268 \text{ K}$$

From equation (6.8)

$$COP = \frac{T_L}{T_H - T_L} = \frac{268}{308 - 268} = 6.7$$

From equation (7.5)

$$COP = \frac{q_L}{q_H - q_L}$$

From equation (5.6), the heat transferred in the condenser can be expressed in terms of the change in enthalpy:

$$q_H = h_3 - h_2$$

This gives a negative value, in accordance with the sign convention that heat rejection is negative. However, to ensure that COP is positive, q_H is re-calculated from:

$$q_H = h_2 - h_3$$

From Appendix A3,

$$h_2 = 201.45 \text{ kJ/kg}, h_3 = 69.55 \text{ kJ/kg}$$

Therefore

$$q_H = 201.45 - 69.55$$
$$= 131.9 \text{ kJ/kg}$$

Substituting in equation (7.5)

$$6.7 = \frac{q_L}{131.9 - q_L}$$

Re-arranging,

$$131.9 - q_L = \frac{q_L}{6.7}$$

Therefore,

$$q_L = \frac{131.9}{1 + \frac{1}{6.7}} = 114.77 \text{ kJ/kg}$$

7.7   THE VAPOUR-COMPRESSION CYCLE

The reversed Carnot cycle is not used in practical refrigeration systems because of the difficulty in achieving an expansion from saturated liquid, at state 3, to a wet vapour at state 4 and gaining a useful work output. For simplicity, it is possible to replace the expander by a throttle valve, as described in section

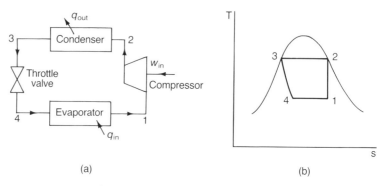

Figure 7.9 Vapour-compression cycle.

5.5.3, to reduce the pressure of the fluid from that in the condenser down to that in the evaporator. Such a system is shown in Figure 7.9(a) and results in the vapour-compression cycle shown in Figure 7.9(b).

Since the flow through a throttle valve is **irreversible**, the expansion is accompanied by an **increase in entropy**, as shown in Figure 7.9(b). The ideal vapour-compression cycle can be considered to consist of four processes:

1–2, reversible adiabatic compression of the vapour;
2–3, isothermal heat rejection to the surroundings;
3–4, irreversible expansion through a throttle valve;
4–1, isothermal heat input from the refrigeration space.

Because the expansion is irreversible the **whole cycle is irreversible**.

The coefficient of performance for the vapour-compression cycle can be defined as

$$COP = \frac{q_{in}}{w_{net}} = \frac{q_{41}}{w_{12}}$$

Expressing these values in terms of the enthalpy changes,

$$q_{41} = h_1 - h_4 \tag{7.6}$$

The net work input is given by

$$w_{12} = h_2 - h_1$$

and combining gives the coefficient of performance:

$$COP = \frac{h_1 - h_4}{h_2 - h_1} \tag{7.7}$$

From equation (5.8), the change in enthalpy across a throttle valve is zero and the value of h_4 can be found from

$$h_4 = h_3$$

Example 7.5

A refrigerator operates on a vapour-compression cycle with Refrigerant-12 as the working fluid, between an evaporator temperature of –5°C and a condenser temperature of 35°C. Calculate the coefficient of performance and the heat transferred to the evaporator.

Conceptual model – Figure 7.9(a).

Process diagram – Figure 7.9(b).

Analysis – the coefficient of performance is defined by equation (7.7):

$$COP = \frac{h_1 - h_4}{h_2 - h_1}$$

but

$$h_4 = h_3$$

so that

$$COP = \frac{h_1 - h_3}{h_2 - h_1}$$

At state 2, from Appendix A3,

$$h_2 = 201.45 \text{ kJ/kg}, \ s_2 = 0.6839 \text{ kJ/kg K}$$

At state 1

$$s_1 = s_2 = 0.6839 \text{ kJ/kg K}$$

Therefore

$$0.6839 = s_f + x_1 (s_g - s_f)$$
$$= 0.1251 + x_1 (0.6991 - 0.1251)$$
$$x_1 = 0.974$$

Taking values of specific enthalpy from Appendix A3,

$$h_1 = h_f + x_1 (h_g - h_f)$$
$$= 31.45 + 0.974 (185.38 - 31.45)$$
$$= 181.38 \text{ kJ/kg}$$

At state 3, from Appendix A3,

$$h_3 = 69.55 \text{ kJ/kg}$$

Therefore

$$COP = \frac{181.38 - 69.55}{201.45 - 181.38} = 5.57$$

The heat transferred to the evaporator is given by equation (7.6):

$$q_{in} = h_1 - h_4 = 181.38 - 69.55$$
$$= 111.83 \text{ kJ/kg}$$

Note – compared to the reversed Carnot cycle working with the same refrigerant, analysed in example 7.4, there is a significant decrease in the coefficient of performance, from 6.7 down to 5.57, but only a slight decrease in the heat transferred in the evaporator, from 114.77 kJ/kg down to 111.83 kJ/kg.

7.8 REFRIGERANTS

There are several fluorocarbon refrigerants that have been developed for use in vapour-compression cycles. Which is the most suitable depends on the temperatures in the cycle. For a domestic storage refrigerator, a temperature slightly above freezing must be maintained in the cabinet. To ensure heat transfer from the cabinet to the evaporator requires a temperature difference. Therefore, the temperature of the evaporator must be some 10 K below the cabinet temperature. In the case of a cabinet temperature of 5°C, the evaporator must operate at a maximum of –5°C.

To ensure heat transfer from the condenser to the surroundings, there must be a temperature difference of 10 K, or more. For a high ambient temperature of, say, 25°C the condenser would be required to operate at a **minimum** of 35°C.

The minimum temperature and, therefore, the minimum pressure in the cycle is achieved in the evaporator. This pressure should be above atmospheric to prevent leakage of air into the system. The maximum temperature and, therefore, the maximum pressure achieved in the cycle is in the condenser. This must be well below the critical point of the refrigerant to ensure isothermal heat rejection. These two requirements define the most suitable refrigerant for a given application. In addition, the refrigerant should not be toxic, corrosive or inflammable if any leakage takes place.

Three widely used fluorocarbon refrigerants are Refrigerant-11, Refrigerant-12 and Refrigerant-22. These are all marketed in the UK by I.C.I. using the trade name 'Arcton'. (In the USA the trade name is 'Freon'.) The characteristics are defined in Table 7.1 together with the characteristics of Refrigerant-134a.

Refrigerant-11 is used for refrigeration system at the upper temperature levels, typically in water chillers and air-conditioning plant. Refrigerant-22 is used where lower temperatures than those readily attained by refrigerant-12 are required or where space is limited as in packaged air-conditioning units. Of the three fluorocarbon refrigerants, Refrigerant-12 is most widely used due to

Table 7.1 Refrigerant characteristics

Refrigerant	11	12	22	134a
Chemical formula	CCl_3F	CCl_2F_2	$CHClF_2$	CH_2FCF_3
Boiling point °C	23.8	−29.8	−40.8	−26.2
Critical temperature °C	198	112	96	100.6
Critical pressure MPa	4.4	4.1	5.0	3.8
Ozone depletion potential	1	1	0.05	0

excellent thermal properties combined with stable chemical characteristics. Properties of Refrigerant-12 are given in Appendix A3.

However, all these three fluorocarbon refrigerants have potential to damage the environment if released into the atmosphere. In particular, they can cause depletion of the ozone layer. To prevent this, a new alternative fluorocarbon, Refrigerant-134a, has been developed and went into production in 1991. From the table above it will be seen that the significant difference between 134a and the other fluorocarbons is the lack of chlorine in its chemical structure. Since the thermal properties are similar, Refrigerant-134a is seen as a replacement for refrigerant-12. The saturated properties of Refrigerant-134a are given in Appendix A4.

Example 7.6

Calculate the coefficient of performance and the heat transferred to the evaporator for the vapour-compression cycle defined in example 7.5, using Refrigerant-134a as the working fluid.

Conceptual model – Figure 7.9(a).

Process diagram – Figure 7.9(b).

Analysis – from example 7.5, the coefficient of performance has been defined as

$$COP = \frac{h_1 - h_3}{h_2 - h_1}$$

At state 2, from Appendix A4,

$$h_2 = 265.53 \text{ kJ/kg}, \, s_2 = 0.9051 \text{ kJ/kg K}$$

At state 1,

$$s_1 = s_2 = 0.9051 \text{ kJ/kg K}$$

Therefore,

$$\begin{aligned} 0.9051 &= s_f + x_1 (s_g - s_f) \\ &= 0.1773 + x_1 (0.9168 - 0.1773) \\ x_1 &= 0.984 \end{aligned}$$

Taking values of specific enthalpy from Appendix A4,

$$\begin{aligned} h_1 &= h_f + x_1 (h_g - h_f) \\ &= 44.52 + 0.984 (242.83 - 44.52) \\ &= 239.66 \text{ kJ/kg} \end{aligned}$$

At state 3, from Appendix A4,

$$h_3 = 99.21 \text{ kJ/kg}$$

Therefore,

$$COP = \frac{239.66 - 99.21}{265.53 - 239.66} = 5.43$$

The heat transferred to the evaporator is given by equation (7.6):

$$q_{in} = h_1 - h_4 = 239.66 - 99.21$$
$$= 140.45 \text{ kJ/kg}$$

Note – comparing the same vapour-compression cycle using Refrigerant-134a instead of Refrigerant-12, there is a slight drop in the coefficient of performance from 5.57 to 5.43, but a significant increase in the heat transferred in the evaporator, from 111.83 kJ/kg up to 140.45 kJ/kg. This means that the mass flow rate for Refrigerant-134a would be less for a given heat load.

7.9   PRACTICAL REFRIGERATOR SYSTEMS

The vapour-compression cycle is the basis for most of the domestic refrigerators and freezers used in practice. A glance at a refrigerator reveals a heat exchanger inside the cabinet which incorporates the evaporator. Behind the refrigerator is a coil, open to the atmosphere, which serves as the condenser and transfers heat to the surroundings. The compressor is generally of a reciprocating type, housed with an electric motor, inside a fully hermetically sealed cabinet.

In practice, refrigerators and freezers work on a vapour-compression cycle that is rather different to the ideal cycle shown in Figure 7.9(b). It is difficult to finish the evaporation precisely at state 1 so that the reversible adiabatic compression will result in saturated vapour at state 2. In practice, evaporation continues until the vapour reaches saturated state or is slightly superheated. During condensation, the cycle performance is improved if the vapour is slightly sub-cooled at state 3 to below the saturation temperature. The modified vapour-compression cycle is shown in Figure 7.10. Finally, the compression will involve some friction and the actual process will thus be irreversible, 1 to 2a, instead of 1 to 2. The analysis of this more complex cycle is beyond the scope of this book, but can be found in more advanced texts. When taking into account all these modifications, the actual coefficient of performance of a vapour-compression system is between 3 and 4.

Not all refrigerators operate on the vapour-compression cycle. Some refrigerators operate by burning gas, either bottled gas or mains gas, and can still be found in the domestic environment or in caravans. Such refrigerators operate on the absorption principle. If a refrigerant can be absorbed into a liquid, the work input required to pump a liquid from the low-pressure evaporator into the higher-pressure condenser, is very much smaller than the work required for a compressor. The basis of an absorption refrigerator system is shown in Figure 7.11.

It will be noticed that the system shown in Figure 7.11 looks very similar to a vapour-compression system except that the compressor has been replaced by a sub-system consisting of an absorber, pump, generator and return valve. The refrigerant is absorbed into a liquid within the absorber, the mixture is then pumped into the generator. Here, the refrigerant is released from the liquid by the transfer of heat from the surroundings. The refrigerant then enters the con-

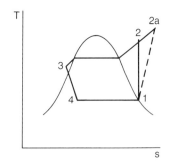

Figure 7.10 A modified vapour-compression cycle.

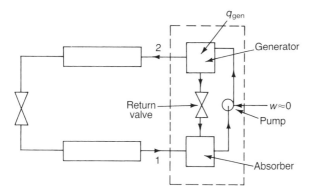

Figure 7.11 Absorption refrigeration system.

denser at state 2, while the liquid returns to the absorber by means of the valve. The most widely used absorption system uses ammonia as the refrigerant and water as the absorbing liquid. For most small refrigerators working on the absorption principle, the actual work input is zero as the mixture is pumped into the generator using a 'bubble-pump' with bubbles created by combining hydrogen gas in the system.

Other absorption refrigeration systems employ water-lithium bromide, where the water serves as the refrigerant. The minimum temperature must be maintained above freezing but such systems are used for air-conditioning, especially where the heat input to the generator can be obtained from a low quality but cheap energy source, i.e. solar energy.

Because of the high heat input required for the generator, the coefficient of performance of an absorption system is lower than for a vapour-compression system. A typical value is about 1 for a small domestic absorption refrigerator.

SUMMARY

In this chapter the application of vapour cycles to both power production and refrigeration has been discussed. The key terms that have been introduced are:

steam power cycles
Carnot steam cycle
Rankine steam cycle
Rankine cycle with superheat
Rankine cycle with two-stage expansion
reversed Carnot cycle
coefficient of performance
vapour compression cycle
fluorocarbon refrigerants
alternative Refrigerant-134a
absorption refrigerant system

Key equations that have been introduced.

For steam power cycles:

$$w_{\text{net}} = \eta\,(h_2 - h_1) \tag{7.1}$$

For the Rankine steam cycle in which the pump work is negligible:

$$h_1 \approx h_4$$

and

$$\eta \approx \frac{h_2 - h_3}{h_2 - h_4} \tag{7.4}$$

For the vapour-compression cycle as applied to a refrigeration system, the heat transferred in the evaporator is

$$q_{\text{in}} = h_1 - h_4 \tag{7.6}$$

and

$$COP = \frac{h_1 - h_4}{h_2 - h_1} \tag{7.7}$$

where the flow across the throttle valve is defined by

$$h_4 = h_3$$

PROBLEMS

Power cycles

1. A steam power plant operates with the following cycle

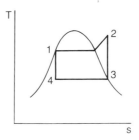

Compared to a Carnot cycle operating between temperatures T_1 and T_4, will there be an improvement or reduction in (a) the thermal efficiency, (b) the net work output?

2. A solar 'powered' engine operates on a vapour cycle with a boiler temperature of 65°C and a condenser temperature of 25°C. If the actual thermal efficiency is half the Carnot efficiency, find the rate of heat transfer to the boiler if the power output of the engine is 100 kW.

3. A steam power plant operates on a Rankine cycle with a boiler pressure of 800 kPa and a condenser pressure of 40 kPa. If the steam leaves the boiler as a saturated vapour, calculate the thermal efficiency of the plant. Ignore the pump work input.

4. A power station operates on a Rankine cycle with steam leaving the boiler at 1 MPa, 400°C, and entering the condenser at 10 kPa. If the power generated is 50 MW, find the rate of heat rejected in the condenser. Ignore the feed pump work in the analysis.

5. A coal fired power station operates on a Rankine cycle in which steam enters a turbine at 2 MPa, 300°C, and exhausts at 20 kPa. If the station has a power output of 200 MW, calculate the required coal consumption in tonnes /hr if the coal has an energy content of 30 000 kJ/kg. Take 1 tonne = 10^3 kg and ignore the feed pump work.

Refrigeration cycles

6. A refrigerator operates on a reversed Carnot cycle with an evaporator temperature of −10°C and a condenser temperature of 30°C. If the rate of heat transfer of the refrigerator is 250 W, calculate the power input.

7. A refrigerator operates on a vapour-compression cycle with Refrigerant-134a as the working fluid. Find the coefficient of performance if the evaporator temperature is −5°C and the condenser temperature is 35°C.

8. A freezer operates on a vapour-compression cycle with Refrigerant-12 as the working fluid. It is required to maintain a cabinet temperature of −10°C with an ambient temperature of 25°C. Assuming a temperature difference of 10 K during the heat transfer processes, calculate the coefficient of performance.

9. A refrigerator operates on a vapour-compression cycle with Refrigerant-134a as the working fluid. Find the coefficient of performance if the cabinet temperature is 0°C and the ambient temperature is 20°C. Assume a 10 K temperature difference for the two heat transfer processes.

    Also, find the mass flow rate of the refrigerant if the rate of heat transfer to the refrigerator is 200 W.

8 Gas power cycles

8.1 AIMS

- To introduce the air-standard cycle as a basis for analysing internal combustion engines.
- To discuss the Carnot cycle as a constant temperature air-standard cycle.
- To outline the practical difficulties in building an engine to work on the Carnot cycle.
- To discuss the Otto cycle as a constant volume air-standard cycle.
- To discuss the influence of the compression ratio on thermal efficiency for the Otto cycle.
- To outline the limitations of the Otto cycle for analysing actual petrol engines.
- To discuss the constant pressure cycle as a basis for analysing gas turbine engines.
- To evaluate the thermal efficiency for a closed-cycle gas turbine engine.
- To discuss the influence of pressure ratio on the performance of a gas turbine engine.
- To discuss the influence of turbine inlet temperature on the performance of a gas turbine engine.
- To model the open-cycle gas turbine engine using the air-standard cycle.

8.2 AIR-STANDARD CYCLES

As the name implies, a gas cycle is one in which the working fluid is a gas. Power cycles that operate with gas as the working fluid are used in reciprocating petrol engines and rotating gas turbine engines. The gas used as the working fluid within these engines is air. This is because air is readily available from the atmosphere and, also, the oxygen in the air supports 'internal' combustion in the engine for the heat input to the thermodynamic cycle.

Use of air within engines means that the air is taken into the engine at the start of a cycle and is then exhausted back to the atmosphere at the end of the cycle. This is termed as 'open-cycle'. This represents one of the main differences between the steam power cycles, considered in the previous chapter, and power cycles operating on air. Modern steam power plants operate on closed cycles, whereas internal combustion engines operate on open cycles. In addition, the combustion of fuel in a steam plant is 'external' to the boiler, whereas engines operating on air have the combustion inside the engine.

These differences make the analysis of an internal combustion engine quite complex. In order to simplify the analysis, certain assumptions have to be made to mathematically model the situation. The simplified cycles used to model the behaviour of engines operating on air are called the 'air-standard cycles'.

The basic requirements for modelling any thermodynamic situation are that the working fluid is a pure substance, that it remains in equilibrium at all times, and undergoes idealized processes within a thermodynamic system. To ensure that these requirements are met, an air-standard cycle is based upon the following assumptions.

1. The working fluid is air, which is considered to be a perfect gas.
2. The air operates under equilibrium conditions throughout the whole cycle.
3. All the processes that make up the cycle are reversible.
4. The internal combustion process is modelled as a heat addition process from the surroundings to the system.
5. The exhaust process is modelled as a heat rejection process from the system to the surroundings.

How realistic these assumptions are, determines how valid the analysis of a particular cycle is with respect to the performance of an actual engine.

As a basis for studying internal combustion engines, three types of air-standard cycle are considered within this chapter. For each of the air-standard cycles the heat addition and heat rejection processes are assumed to take place under the same conditions. The three types of air-standard cycle to be analysed are:

constant temperature cycle
constant volume cycle
constant pressure cycle

For each of these cycles the main performance criterion is taken to be the thermal efficiency of the cycle.

8.3 CONSTANT TEMPERATURE CYCLE

A cycle in which both the heat addition and heat rejection processes are at constant temperature, i.e. isothermal, is the Carnot cycle. The performance of the Carnot cycle is independent of the working fluid. Therefore, Carnot cycle operating on air, or any other perfect gas, would have the same thermal efficiency as a Carnot cycle operating on a vapour between the same temperature limits.

In theory, the Carnot cycle operating with air as a working fluid should have a higher temperature than steam for the heat input although; in practice, this will be limited by the materials used in the construction of any such engine.

The T–s process diagram for the Carnot air cycle is shown in Figure 8.1. It consists of four processes:

1–2, isothermal heat addition;
2–3, reversible adiabatic expansion;
3–4, isothermal heat rejection;
4–1, reversible adiabatic compression.

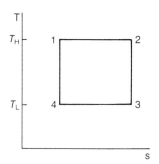

Figure 8.1 The Carnot air cycle.

The thermal efficiency for such a cycle has already been defined by equation (6.6):

$$\eta = 1 - \frac{T_L}{T_H}$$

Defining the Carnot cycle as an ideal air-standard cycle is easy. Defining a thermodynamic system that would allow it to be achieved in practice is very much more difficult. If the cycle is assumed to take place inside a closed system consisting of a cylinder and frictionless piston, the heat transfer processes across the boundary have to change for each process. To achieve isothermal heat input to a gas, the gas must be expanding while the process takes place slowly in order for heat to be transferred between the surroundings and the system. During the adiabatic expansion, either the cylinder must be suddenly insulated (which is impractical) or the process must be extremely rapid so that heat transfer to the surroundings is negligible.

Figure 8.2 shows the variation of the pressure and specific volume of air inside a cylinder and piston, while undergoing a Carnot cycle. In order to achieve such a cycle, the velocity at which the piston moves should be slow from state 1 to 2 and then rapid from 2 to 3. The reverse stroke then involves slow movement from 3 to 4 and rapid movement from 4 to 1. Such differential movement could be achieved, although the resulting engine would be complicated. However, the heat transfer conditions across the boundary would be far more difficult to achieve. The reason why such an engine has not been built in practice can be found from the P–v diagram in Figure 8.2. It was shown in section 3.5.3 that the work output from a cycle is the enclosed area on the P–v diagram. The enclosed area of the P–v diagram in Figure 8.2 can be seen to be small, indicating that the specific work output from such an engine would be very low.

Before leaving this discussion of the Carnot air cycle, it is instructive to look back at the Carnot steam cycle portrayed in Figure 7.2. The system proposed for the Carnot steam cycle involved steady flow devices. This raises the obvious question as to whether a system could be designed, using steady flow devices, that would allow the Carnot air cycle to be achieved in practice.

An engine involving steady flow devices and operating on a Carnot cycle is shown in Figure 8.3. Assuming the working fluid to be air, although any gas could be used with a closed system of this nature, the engine operates with two turbines and two compressors. Air enters the first turbine at state 1 and undergoes an isothermal expansion with heat transfer from the surroundings to

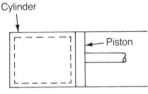

Cylinder

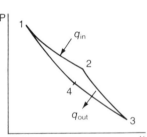

Figure 8.2 A Carnot air cycle within a cylinder and piston.

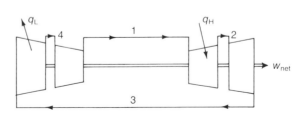

Figure 8.3 A Carnot steady flow engine.

ensure that the temperature remains constant. The air leaves the isothermal turbine at state 2 and expands adiabatically to state 3. It then enters an isothermal compressor to transfer heat from the air to the surroundings. Finally, the air is compressed adiabatically from state 4 to state 1.

Although a more feasible type of engine than that proposed in Figure 8.2 the steady flow engine is not a practical proposition. It would be difficult to achieve isothermal processes in a turbine and compressor. Even if achieved, the friction in the individual components would mean that the actual performance would fall far short of the ideal portrayed by the Carnot cycle.

The usefulness of the Carnot cycle lies in the fact that it is the most efficient thermodynamic cycle and, therefore, provides a yardstick by which other cycles can be judged. It also provides a guide by which the efficiency of an air-standard cycle or an engine can be improved. Thermal efficiency of an engine improves if the average temperature at which heat is added to the system is increased. This assumes that the average temperature at which heat is rejected from the system is governed by atmospheric conditions and cannot be changed. Therefore, the improvement in thermal efficiency of internal combustion engines in this century has been achieved by increasing maximum temperatures, made possible by improvements in the materials used within the engines.

Example 8.1
Calculate the thermal efficiency of an engine, with air as the working fluid, operating between a maximum temperature of 1200 K and an ambient temperature of 20°C. Assume the engine operates on a Carnot cycle.

Conceptual model – as outlined in the foregoing discussion an engine operating on a Carnot cycle can be modelled using Figure 8.3.

Process diagram – Figure 8.1.

Analysis

$$T_H = 1200 \text{ K}$$
$$T_L = 20 + 273 = 293 \text{ K}$$

From equation (6.6)

$$\eta = 1 - \frac{T_L}{T_H} = 1 - \frac{293}{1200} = 0.756 \text{ i.e. } 75.6\%$$

8.4   CONSTANT VOLUME CYCLE

A cycle in which both the heat addition and heat rejection take place at constant volume is called the 'Otto' cycle, after Nikolaus Otto, and is the basis for analysing spark ignition engines.

8.4.1   The Otto cycle

Using the air-standard assumptions the Otto cycle can be visualized as shown in Figure 8.4. Assuming that the cycle takes place inside a closed system

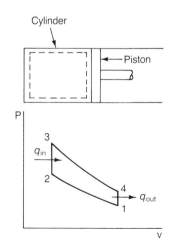

Figure 8.4 P–v diagram of the constant volume cycle.

consisting of a cylinder and frictionless piston, the cycle consists of four processes:

1–2, reversible adiabatic compression;
2–3, heat addition at constant volume;
3–4, reversible adiabatic expansion;
4–1, heat rejection at constant volume.

The changes in pressure and specific volume that take place within the cycle are shown in the P–v diagram given in Figure 8.4. This clearly illustrates the effect of transferring heat at constant volume for both the heat addition and heat rejection processes.

For a constant volume process, the heat transferred to and from the working fluid can be expressed as

$$q_{in} = q_{23} = C_v \, (T_3 - T_2)$$
$$q_{out} = -q_{41} = C_v \, (T_4 - T_1)$$

The net work output of the cycle is

$$w_{net} = q_{in} - q_{out}$$
$$= C_v \, (T_3 - T_2) - C_v \, (T_4 - T_1)$$

Dividing the net work output by the heat input gives the thermal efficiency:

$$\eta = \frac{C_v \, (T_3 - T_2) - C_v \, (T_4 - T_1)}{C_v \, (T_3 - T_2)}$$

Air is assumed to be a perfect gas, so that C_v is taken to be constant and the thermal efficiency can be expressed:

$$\eta = 1 - \frac{(T_4 - T_1)}{(T_3 - T_2)} \tag{8.1}$$

For the adiabatic processes 1–2 and 3–4, the variation of temperature with specific volume is given by equation (4.24):

$$\frac{T_2}{T_1} = \left(\frac{v_1}{v_2} \right)^{\gamma - 1}$$

and

$$\frac{T_3}{T_4} = \left(\frac{v_4}{v_3} \right)^{\gamma - 1}$$

However, the ratio of the maximum volume to the minimum volume is constant:

$$\frac{v_1}{v_2} = \frac{v_4}{v_3} = r \tag{8.2}$$

where r is the 'compression ratio' for the cycle.

Example 8.2

Calculate the thermal efficiency of a petrol engine operating on an Otto air-standard cycle between a maximum temperature of 1200 K and an ambient temperature of 20°C, if the compression ratio is 8. Take $\gamma = 1.4$ for air.

Conceptual model – Figure 8.4.

Process diagram – Figure 8.4.

Analysis

$$T_1 = 20 + 273 = 293 \text{ K}$$

From equation (4.24)

$$T_2 = T_1 \left(\frac{v_1}{v_2} \right)^{\gamma - 1}$$

From equation (8.2)

$$\frac{v_1}{v_2} = r = 8$$

Therefore,

$$T_2 = 293(8)^{0.4} = 673.1 \text{ K}$$

Similarly,

$$T_4 = 1200(1/8)^{0.4} = 522.3 \text{ K}$$

Substituting in equation 8.1

$$\eta = 1 - \frac{(T_4 - T_1)}{(T_3 - T_2)}$$

$$= 1 - \frac{(522.3 - 293)}{(1200 - 673.1)} = 0.565 \quad \text{i.e. } 56.5\%$$

Note – A comparison with the thermal efficiency for the Carnot cycle operating between the same temperature limits, shows a noticeable drop from 75.6% down to 56.5%.

8.4.2 Effect of compression ratio

What is not obvious from example 8.2 is that the thermal efficiency would have been the same irrespective of what maximum temperature was chosen, providing that T_3 was above 673.1 K. This is because the ratio of the temperature differences

$$(T_4 - T_1)/(T_3 - T_2)$$

is the same for any given compression ratio.

It can be shown that the ratio of the temperature differences is equal to the ratio T_1/T_2, i.e.

$$\frac{(T_4 - T_1)}{(T_3 - T_2)} = \frac{T_1}{T_2}$$

This can be illustrated taking numerical values from example 8.2:

$$\frac{(T_4 - T_1)}{(T_3 - T_2)} = 0.435$$

and

$$\frac{T_1}{T_2} = \frac{293}{673.1} = 0.435$$

Substituting the ratio T_1/T_2 in equation (8.1) gives

$$\eta = 1 - \frac{T_1}{T_2}$$

but,

$$T_2 = T_1 \, r^{\gamma-1}$$

so that the thermal efficiency of an Otto air-standard cycle is given by

$$\eta = 1 - \frac{1}{r^{\gamma-1}} \tag{8.3}$$

The variation of thermal efficiency with compression ratio for the Otto air-standard cycle is shown in Figure 8.5. This clearly shows that thermal efficiency improves as the compression ratio is increased and this trend is true for

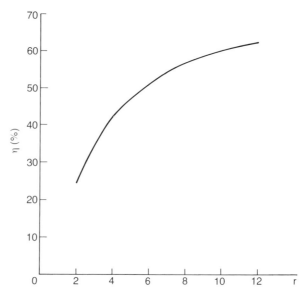

Figure 8.5 Variation of thermal efficiency with compression ratio.

real spark-ignition engines. However, the improvement tends to flatten out for compression ratios above 8.

In practice, there is a limit to which compression ratios can be increased. In a real engine the compression process, from state 1 to state 2, involves not just air but a mixture of air and petrol. If the temperature of the mixture is increased too much, the mixture will self-ignite. This self-ignition, called detonation, causes a reduction in the performance of the engine and, if it continues, mechanical damage. Detonation is accompanied by an audible knocking noise, colloquially known as 'pinking'.

Improvement of the thermal efficiency of petrol engines has been made possible by increasing compression ratios as a result of using petrols with good anti-knock properties. Figure 8.6 shows the general increase in the compression ratio of car engines.

The anti-knock properties are determined by the 'octane rating' of the petrol. Octane is a liquid hydrocarbon fuel with the chemical formula C_8H_{18}. The octane rating gives the anti-knock properties equivalent to a percentage of octane in a mixture of octane and heptane. For example, an octane rating of 100 would have the same anti-knock properties as a fuel containing 100% octane. Typical octane-ratings for particular compression ratios are:

Compression ratio	Octane rating
6.8	75
8.2	100

The cheapest way of improving the octane rating is by adding tetraethyl lead to the petrol. Leaded petrol was first developed in the 1920s and resulted in the significant increase in compression ratios shown in Figure 8.6. Use of lead can result in the anomolous situation in which petrols have been developed with octane ratings greater than 100. Unfortunately, leaded petrols have a detrimental effect on health and the environment. Since the 1980s there has been a move to unleaded petrols in which the anti-knock properties are enhanced by mixing alcohol in the petrol, a more expensive way of improving the octane rating than tetraethyl lead.

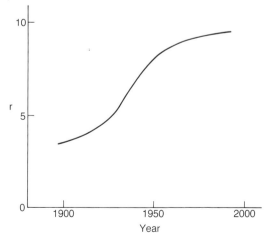

Figure 8.6 Development of compression ratio for car engines.

Example 8.3

A petrol engine operates on an Otto air-standard cycle with a compression ratio of 8. If the engine produces a steady power output of 30 kW calculate the petrol consumption if the fuel has an energy content of 44 MJ/kg. Assume $\gamma = 1.4$ for air.

Conceptual model – Figure 8.4.

Process diagram – Figure 8.4.

Analysis – the thermal efficiency of the cycle can be found from equation (8.3)

$$\eta = 1 - \frac{1}{r^{\gamma-1}} = 1 - \frac{1}{8^{0.4}} = 0.565$$

From equation (6.2)

$$\eta = \frac{W_{out}}{Q_{in}}$$

Therefore,

$$Q_{in} = 30/0.565 = 53.1 \text{ kW}$$

To provide this rate of heat input, the engine must use petrol at the rate of

$$\dot{m}_f = \frac{53.1 \times 10^3}{44 \times 10^6} = 0.00121 \text{ kg/s}$$

Note – If petrol has a specific volume of 0.0012 m3/kg, the volume flow rate is equal to 5.2 l/hr.

8.4.3 Limitations of the Otto air-standard cycle

A real petrol engine with a compression ratio of 8, would have a **much** lower thermal efficiency than the value of 56.5% calculated in example 8.3. This is because of the limitations of the Otto air-standard cycle as a way of modelling the behaviour of a real spark-ignition engine.

The idealized cycle is based upon the assumption that the compression and expansion processes are adiabatic. In practice, there is heat transfer to the cylinder wall which requires the engine to have a cooling system. To ensure that temperatures within the engine components are maintained at reasonable levels, cooling water flows around each cylinder of an engine and is itself cooled in a radiator. This represents a significant loss of energy from the cycle.

More importantly, the air-standard cycle is based upon a fixed mass of air within the cylinder undergoing idealized processes. In practice, an engine takes in air from the atmosphere and exhausts the combustion gases back to the atmosphere. This involves work being done to get the air into the cylinder and work being done getting the exhaust gases out of the cylinder. Also, the heat input is achieved by internal combustion which requires a finite time period, not instantaneous heat input as assumed in the constant volume cycle. A

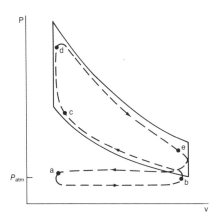

Figure 8.7 Comparison between the air-standard cycle and a real cycle.

comparison between the air-standard cycle and a real cycle is shown in Figure 8.7. The air-standard cycle is shown in full lines, whereas the real cycle is shown in dotted lines.

The real cycle shown in Figure 8.7 is typical of that for a four-stroke engine. The inlet valve opens at 'a' and an air-petrol mixture is drawn into the cylinder. To achieve this, the pressure in the cylinder must be below atmospheric pressure. The inlet valve closes at 'b' and the mixture is compressed to 'c', at which point combustion starts. Combustion ends at 'd' and the hot gases expand to 'e', at which point the exhaust valve opens. The spent gases are then exhausted to atmosphere, but to achieve this the pressure in the cylinder must be above atmosphere.

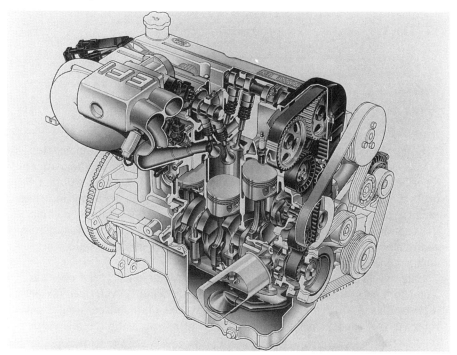

Figure 8.8 A 16-valve petrol engine by Ford.

One way of improving both the thermal efficiency and the net work output is to reduce the work lost during the inlet and exhaust strokes, e–a–b on the diagram. This is why so many manufacturers now use four valves per cylinder, two inlet and two exhaust, to minimize the pressure drop between the cylinder and atmosphere. Such an engine is shown in Figure 8.8 and makes an interesting comparison with that shown in Figure 1.5.

8.5   CONSTANT PRESSURE CYCLE

A cycle in which both the heat addition and heat rejection take place at constant pressure is shown in Figure 8.9. The cycle shown in Figure 8.9 consists of four processes:

1–2, reversible adiabatic compression;
2–3, heat addition at constant pressure;
3–4, reversible adiabatic expansion;
4–1, heat rejection at constant pressure.

As such, the constant pressure cycle is not applicable to closed systems consisting of a cylinder and piston. Instead, it is the basis of the gas turbine engine.

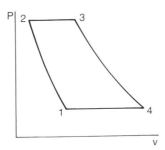

Figure 8.9 P–v diagram of the constant pressure cycle.

Some textbooks refer to the constant pressure cycle as the Joule cycle. This is because in 1851, Joule proposed a form of hot air engine in which air would flow between two large chambers. A hot chamber, heated by external combustion, and a cold chamber. The air would flow between the two using either a reciprocating compressor or a reciprocating expander. Assuming the volume of the chambers to be large in comparison to the reciprocators, the pressure in each chamber would be virtually constant and the engine would work on a constant pressure cycle.

Later, in 1873, Brayton built the engine described in Chapter 1 operating on a constant pressure cycle. As a result others, particularly American textbooks, refer to it as the Brayton cycle. To avoid any bias in this discussion, it will be referred to as the gas turbine cycle because of its application to this particular type of engine.

8.5.1 Closed gas turbine cycle

Before going on to discuss the open-cycle gas turbine engine, it is necessary to see how the constant pressure air-standard cycle shown in Figure 8.9 can be achieved.

Figure 8.10(a) shows a closed-cycle gas turbine engine which consists of four components:

1–2, an adiabatic compressor;
2–3, a heat exchanger in which heat is transferred from the surroundings to the
    working fluid at constant pressure;
3–4, an adiabatic turbine;
4–1, a heat exchanger in which heat is rejected to the surroundings at constant
    pressure.

Because the closed-cycle gas turbine engine operates with steady flow devices, the most appropriate process diagram for the cycle is the T–s diagram shown in Figure 8.10(b). The heat addition, 2 to 3, and heat rejection, 4 to 1, processes are represented on lines of constant pressure. The compression and expansion processes are assumed to be reversible adiabatic processes with zero change of entropy.

The thermal efficiency for the cycle can be evaluated from

$$\eta = \frac{w_{net}}{q_{in}}$$

The net work output of the cycle is the sum of the work from the turbine and compressor:

$$w_{net} = w_{turb} + w_{comp}$$

Since the compressor and turbine are steady flow devices the work can be evaluated from the change of enthalpy, from equation (5.9):

$$w_{net} = (h_3 - h_4) + (h_1 - h_2)$$

However, the working fluid is a gas so the change of enthalpy can be evaluated using equation (4.13):

$$w_{net} = C_P (T_3 - T_4) + C_P (T_1 - T_2)$$

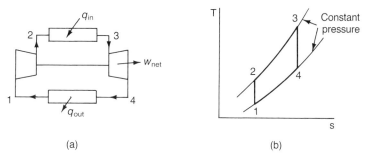

(a)   (b)

Figure 8.10 Closed-cycle gas turbine engine.

Re-arranging,

$$w_{net} = C_P (T_3 - T_4) - C_P (T_2 - T_1) \tag{8.4}$$

The heat input to the cycle can be evaluated in the same way, so that

$$q_{in} = q_{23} = C_P (T_3 - T_2) \tag{8.5}$$

Combining equations (8.4) and (8.5) gives a relationship for the thermal efficiency:

$$\eta = \frac{C_P (T_3 - T_4) - C_P (T_2 - T_1)}{C_P (T_3 - T_2)}$$

Air is assumed to be a perfect gas, so that C_P is taken to be constant and the thermal efficiency expressed as

$$\eta = \frac{T_3 - T_4 - T_2 + T_1}{T_3 - T_2} \tag{8.6}$$

For the adiabatic processes 1–2 and 3–4, the variation of temperature with pressure is given by equation (4.25):

$$\frac{T_2}{T_1} = \left(\frac{P_2}{P_1} \right)^{(\gamma-1)/\gamma}$$

and

$$\frac{T_3}{T_4} = \left(\frac{P_3}{P_4} \right)^{(\gamma-1)/\gamma}$$

However, the pressure ratio is constant for the closed gas turbine cycle:

$$\frac{P_2}{P_1} = \frac{P_3}{P_4}$$

Example 8.4

Calculate the thermal efficiency of a closed-cycle gas turbine engine operating on air between a maximum temperature of 1200 K and a minimum temperature of 20°C. Assume the pressure ratio to be 10 and take $\gamma = 1.4$ for air.

Conceptual model – Figure 8.10(a).

Process diagram – Figure 8.10(b).

Analysis

$$T_1 = 20 + 273 = 293 \text{ K}$$

From equation (4.25)

$$T_2 = T_1 \left(\frac{P_2}{P_1} \right)^{(\gamma-1)/\gamma} = 293(10)^{0.4/1.4} = 565.7 \text{ K}$$

$$T_4 = T_3 \left(\frac{P_4}{P_3} \right)^{(\gamma-1)/\gamma} = \frac{1200}{(10)^{0.4/1.4}} = 621.5 \text{ K}$$

Substituting in equation (8.6)

$$\eta = \frac{T_3 - T_4 - T_2 + T_1}{T_3 - T_2}$$

$$= \frac{1200 - 621.5 - 565.7 + 293}{1200 - 565.7} = 0.482 \text{ i.e. } 48.2\%$$

Note – this value of thermal efficiency can be compared with the values for the Otto cycle and Carnot cycles operating between the same temperature limits, as in examples 8.1 and 8.2.

8.5.2 Effect of pressure ratio

What is not obvious from example 8.4, is that the thermal efficiency would have been the same irrespective of what maximum temperature was chosen, providing that T_3 was above 565.7 K. Just as the thermal efficiency of the constant volume air-standard cycle is a function of the volume ratio, i.e. the compression ratio, so the thermal efficiency of the constant pressure air-standard cycle is a function of the pressure ratio alone. It is not proposed to prove this but a graph showing the improvement of thermal efficiency with pressure ratio is shown in Figure 8.11.

Figure 8.11 clearly shows that thermal efficiency improves as the pressure ratio is increased and this trend is true for real gas turbine engines. However, the improvement tends to flatten out for pressure ratios above 15. In practice, there is a limit to which pressure ratios can be increased, as this depends on compressor technology. For example, the jet engine shown in Figure 1.7 was designed in the early 1940s and employed a centrifugal compressor with a pressure ratio of 4. This represented the limit for the technology at that time. By comparison, modern gas turbine engines have axial-flow compressors

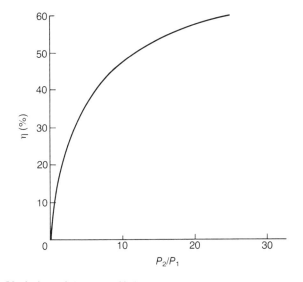

Figure 8.11 Variation of thermal efficiency with pressure ratio.

incorporating a large number of stages, having overall pressure ratios of up to 30.

As well as compressor technology there is a fundamental performance criterion that determines the required pressure ratio of a gas turbine engine, the specific work output of the cycle. This criterion was used with respect to the steam power cycles, discussed in the previous chapter. So far, it has not been mentioned with respect to air-standard cycles within this chapter. This is because it is a criterion that is not particularly relevant to either the Carnot or the Otto air-standard cycles.

The specific work output of both these cycles depends on the specific heat input, which is **not** a fixed quantity. In the case of the Otto air-standard cycle the specific work output depends on the compression ratio and the maximum temperature within the cycle. The maximum temperature is, in practice, determined by the air–fuel ratio of the mixture and the speed at which combustion takes place, both of which depend on the design characteristics of a particular engine.

By comparison, the general behaviour of a gas turbine engine can be defined more precisely. The maximum temperature within the cycle is governed by the material limits of the turbine. In practice, the maximum temperature of a gas turbine engine is **much** lower than that achieved within a reciprocating petrol engine. This is because the gas turbine engine operates with steady flow and the maximum temperature does not vary, whereas the temperature inside a petrol engine **varies throughout the cycle**.

The way that the maximum temperature within a gas turbine cycle is controlled is discussed later in this chapter. For the present it is sufficient to accept that the designer can define the maximum temperature within the cycle.

For a defined inlet temperature T_{min} and maximum temperature T_{max}, the actual gas turbine cycle can vary according to the pressure ratio. Figure 8.12 shows two such cycles operating between the same temperature limits. Cycle 1–2–3–4 has a high pressure ratio, whereas cycle 1–2′–3′–4′ has a low pressure ratio. Between these two cycles there are an infinite number of possible cycles with intermediate pressure ratios. The question that obviously comes to mind is which cycle to choose.

Of the two possible cycles shown in Figure 8.12, 1–2–3–4 has the better thermal efficiency because it operates with the higher pressure ratio. However, it is clear from looking at the two cycles that 1–2′–3′–4′ has the greater area and, therefore, the greater specific work output. The specific work output is important because it determines the mass flow rate of air flowing through the engine for a given power output. This in turn determines the size of the engine.

The size of a gas turbine engine is an important design feature. In the case of an industrial gas turbine engine the size will be a major factor in determining the capital cost. In the case of an aircraft gas turbine engine the size, and weight, of the engine **must** be minimized. Since, for any given gas turbine engine, T_{max} and T_{min} are fixed, the actual pressure ratio chosen must be that giving the best specific work output within these limits.

Taking the specific work output as being the net work output, defined in equation (8.4),

$$w = C_P (T_3 - T_4) - C_P (T_2 - T_1)$$

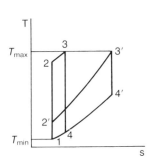

Figure 8.12 Different cycles operating between T_{max} and T_{min}.

it can be shown[*] that the work output is an optimum when

$$T_2 = \sqrt{T_3 \times T_1}$$

$$(8.7)$$

The optimum pressure ratio can then be found from equation (4.25):

$$\frac{P_2}{P_1} = \left(\frac{T_2}{T_1}\right)^{\gamma/(\gamma-1)}$$

Example 8.5
Calculate the specific work output for the closed-cycle gas turbine engine defined in example 8.4. Take C_P for air as 1.005 kJ/kg K.

Conceptual model – Figure 8.10(a)

Process diagram – Figure 8.10(b)

Analysis – taking values from example 8.4,

$$T_1 = 293 \text{ K}$$
$$T_2 = 565.7 \text{ K}$$
$$T_3 = 1200 \text{ K}$$
$$T_4 = 621.5 \text{ K}$$

Substituting in equation 8.4
$$w_{net} = C_P (T_3 - T_4) - C_P (T_2 - T_1)$$

$$= 1.005 (1200 - 621.5) - 1.005 (565.7 - 293)$$
$$= 307.3 \text{ kJ/kg}$$

Example 8.6
Recalculate the specific output and thermal efficiency for the gas turbine engine defined in example 8.4, assuming that the engine operates between the same temperature limits with optimum pressure ratio. For air, assume $C_P = 1.005$ kJ/kg K, $\gamma = 1.4$

---

[*]
$$\frac{P_2}{P_1} = \frac{P_3}{P_4} \quad \text{so that} \quad \frac{T_2}{T_1} = \frac{T_3}{T_4}$$

and

$$T_4 = T_3 \frac{T_1}{T_2}$$

Equation (8.4) then becomes:

$$w = C_P \left(T_3 - T_3\frac{T_1}{T_2} - T_2 + T_1\right)$$

For optimum work output:

$$\frac{dw}{dT_2} = C_P \left(\frac{T_3 T_1}{T_2^{\,2}} - 1\right) = 0$$

$$T_2 = \sqrt{T_3 T_1}$$

Conceptual model – Figure 8.10(a)

Process diagram – Figure 8.10(b)

Analysis – taking values from example 8.4,

$$T_1 = 293 \text{ K}$$
$$T_3 = 1200 \text{ K}$$

For optimum work output, from equation (8.7),

$$T_2 = \sqrt{T_3 T_1} = 1200 \times 293 = 593 \text{ K}$$

The optimum pressure ratio can be found using this value of T in equation (4.25):

$$\frac{P_2}{P_1} = \left(\frac{593}{293}\right)^{1.4/0.4} = 11.8$$

Using this pressure ratio

$$T_4 = T_3 \left(\frac{P_4}{P_3}\right)^{(\gamma-1)/\gamma} = 1200 \left(\frac{1}{11.8}\right)^{0.4/1.4} = 593 \text{ K}$$

Substituting in equation (8.4)

$$w_{\text{net}} = C_P (T_3 - T_4) - C_P (T_2 - T_1)$$
$$= 1.005 (1200 - 593) - 1.005 (593 - 293)$$
$$= 308.5 \text{ kJ/kg}$$

From equation (8.6)

$$\eta = \frac{T_3 - T_4 - T_2 + T_1}{T_3 - T_2}$$
$$= \frac{1200 - 593 - 593 + 293}{1200 - 593} = 0.506 \text{ i.e. } 50.6\%$$

Note – it will be seen that the optimum pressure ratio occurs when T_2 is equal to T_4.

8.5.3 Effect of maximum temperature

Comparing the results from examples 8.5 and 8.6, it will be seen that increasing the pressure ratio from 10 to the optimum value of 11.8 increases the specific work output by only a small amount, 308.5 kJ/kg compared to 307.3 kJ/kg. This indicates that the variation of specific work output is very small at, or near, the value of optimum pressure ratio. The variation of the specific work output with pressure ratio for a maximum temperature of 1200 K, is shown in Figure 8.13.

From Figure 8.13 it can be seen that increasing the pressure ratio above the optimum value has very little effect on the specific work output. It would be possible to choose a pressure ratio of, say, 16 and this would give a negligible drop in specific work output. On the other hand, a pressure ratio of 16 would

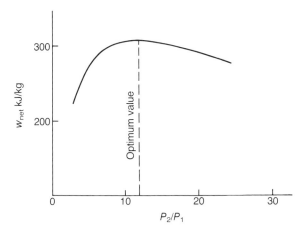

Figure 8.13 Variation of specific work output with pressure ratio.

result in a considerable improvement to the thermal efficiency, as shown in Figure 8.11. Therefore, it would be better design policy to choose an operating pressure ratio of 16, in preference to the optimum value of 11.8.

This is also true for real gas turbine engines, although in practice the actual choice of pressure ratio is more complex because of friction within the compressor and turbine. The air-standard cycle shown in Figure 8.10(b) is based upon the assumption that both the compression, 1–2, and expansion, 3–4, are reversible adiabatic processes. In practice, both these processes would have significant friction, that would reduce the specific work output from the engine.

Nevertheless, the air-standard cycle clearly illustrates the basic design parameters for a gas turbine engine. The pressure ratio is clearly of the greatest importance, but it is, in turn, dependant on the maximum temperature within the cycle. As T_3 is increased the specific work output is increased. The thermal efficiency will also increase because the cycle can operate with a higher pressure ratio.

Gas turbine engine development has been based on improving the maximum operating temperature in the cycle. This is governed by the maximum temperature that the turbine can withstand. Typically, turbine inlet temperatures have increased from about 1000 K in the early 1950s to current values of 1400 K for industrial gas turbine engines. Some aircraft gas turbine engines achieve even higher temperatures and values of 1500 K are quite common. These improvements have been achieved through:

improvements in materials;
turbine blade cooling.

Although improvements in materials have been significant, it would be impossible to operate with material temperatures very much greater than 1000 K. To operate with turbine inlet temperatures greater than this means that the blade temperatures must be maintained within reasonable limits through cooling.

Turbine blade cooling employs a small fraction of the high-pressure air from the outlet of the compressor. The cooling air is fed through channels in

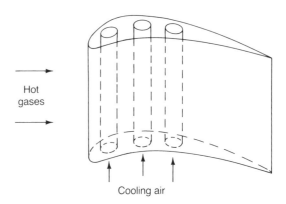

Figure 8.14 Turbine blade cooling.

the blades, as shown in Figure 8.14. This diagram shows a simple form of convective cooling arrangement. If the gas temperature is, say, 1200 K, the cooling air will be at a much lower temperature of perhaps 700 K and the blade material will reach an intermediate temperature. To achieve higher turbine inlet temperatures requires a more complex form of arrangement in which the cooling air is fed through small holes to form a thin film over the surface of the blade.

Example 8.7

An industrial gas turbine engine operates with a pressure ratio of 14 and a turbine inlet temperature of 1200 K. If improvements in blade cooling allow the maximum temperature to be increased to 1300 K, calculate the improvement in the specific work output. Assume the working fluid to be air with a minimum temperature of 20°C, $C_P = 1.005$ kJ/kg K and $\gamma = 1.4$.

Conceptual model – Figure 8.10(a)

Process diagram –

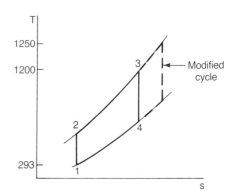

Analysis

$$T_1 = 20 + 273 = 293 \text{ K}$$

For T_3 at 1200 K

$$T_2 = T_1 \left(\frac{P_2}{P_1} \right)^{(\gamma-1)/\gamma} = 293(14)^{0.4/1.4} = 622.8 \text{ K}$$

$$T_4 = T_3 \left(\frac{P_4}{P_3} \right)^{(\gamma-1)/\gamma} = 1200 \left(\frac{1}{14} \right)^{0.4/1.4} = 564.6 \text{ K}$$

$$w_{\text{net}} = C_P (T_3 - T_4) - C_P (T_2 - T_1)$$

$$= 1.005(1200 - 564.6) - 1.005(622.8 - 293)$$

$$= 307.1 \text{ kJ/kg}$$

For T_3 at 1300 K:

$$T_2 = 622.8 \text{ K, from above}$$

$$T_4 = 1300 \left(\frac{1}{14} \right)^{0.4/1.4} = 611.6 \text{ K}$$

$$w_{\text{net}} = 1.005(1300 - 611.6) - 1.005(622.8 - 293)$$

$$= 360.4 \text{ kJ/kg}$$

Note – Clearly the increase in T_3 would allow a significant improvement in the specific work output, from 307.1 kJ/kg to 360.4 kJ/kg, an increase of 17%.

8.5.4 Open-cycle gas turbine engine

It is interesting to compare the specific work output for a gas turbine engine, as given in example 8.7, with the specific work output for a simple steam power cycle, as defined in example 7.2 in the last chapter:

	T_{max}	w_{net}
gas turbine cycle	1300 K	360.4 kJ/kg
steam power cycle	452.9 K	609 kJ/kg

The specific work output for the gas turbine cycle is much lower than for the steam power cycle, and this for a maximum cycle temperature that is much higher. Increasing the superheat temperature for the steam cycle would make the difference even more marked. This difference in the specific work output is why steam is still used in power stations in which high power outputs are required.

Steam generator sets with power outputs of 1500 MW have been built. By comparison, a large industrial gas turbine engine has a power output of 50 MW, although the majority are in the 2 to 25 MW power range. Below a power output of 2 MW it is generally more economic to use a reciprocating diesel engine.

For operation where a closed-cycle is required, for example in nuclear power stations where the working fluid must be contained, steam is a more suitable

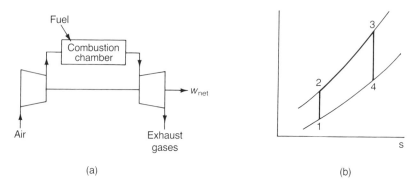

Figure 8.15 Open-cycle gas turbine engine.

working fluid than a gas. Although a closed-air-standard cycle provides the basis for analysing industrial gas turbine engines, it has little application in practice. Actual gas turbine engines operate on an open-cycle system, as shown in Figure 8.15(a).

Air enters the compressor from the atmosphere at state 1 and is compressed to state 2. The air then enters a combustion chamber where it is mixed with fuel and combustion takes place. Finally, the hot gases resulting from combustion flow through the turbine to be rejected to the atmosphere at state 4.

Precise control of the fuel entering the combustion chamber ensures a very close control of the maximum temperature at state 3. With the comparatively low values of T_3 used in engines, the mass flow rate of fuel is very small compared to the air flow. An air to fuel ratio of 50 is typical, so that the mass flow rate of the fuel can be safely ignored in the analysis. As most of the air flows through the combustion chamber without any change, other than an increase in temperature, the combustion gases can be considered to behave in the same way as air. Therefore, the combustion can be modelled as a constant pressure process with the fuel being represented as an external heat input from the surroundings. Using this, together with the assumption that both the compressor inlet and turbine outlet are at the same atmospheric pressure, the cycle can be modelled on the process diagram as shown in Figure 8.15(b).

In all its essential features the open-cycle, shown in Figure 8.15(b) is the same as the closed-cycle shown in Figure 8.10(b) and can be analysed in exactly the same way. Therefore, equation (8.4) for the thermal efficiency and equation (8.6) for the specific work output, can also be applied directly to the open-cycle shown in Figure 8.15(b).

Example 8.8

An industrial gas turbine engine takes in air at 15°C and operates with the following conditions:

mass flow rate = 10 kg/s;
pressure ratio = 12;
turbine inlet temperature = 1000°C.

Calculate the power output and thermal efficiency of the engine. Assume air to have the properties $C_P = 1.005$ kJ/kg K, $\gamma = 1.4$.

Conceptual model – Figure 8.15(a).

Process diagram – Figure 8.15(b).

Analysis

$$T_1 = 15 + 273 = 288 \text{ K}$$
$$T_3 = 1000 + 273 = 1273 \text{ K}$$

From equation (4.25)

$$T_2 = 288(12)^{0.4/1.4} = 585.8 \text{ K}$$

$$T_4 = 1273\left(\frac{1}{12}\right)^{0.4/1.4} = 625.9 \text{ K}$$

Substituting in equation (8.4)

$$
\begin{aligned}
w_{\text{net}} &= C_P(T_3 - T_4) - C_P(T_2 - T_1) \\
&= 1.005(1273 - 625.9) - 1.005(585.8 - 288) \\
&= 351 \text{ kJ/kg}
\end{aligned}
$$

Therefore,

$$
\begin{aligned}
\text{power output} &= \dot{m} \times w_{\text{net}} \\
&= 10 \times 351 = 3510 \text{ kW i.e. 3.5 MW}
\end{aligned}
$$

Substituting the temperature in equation (8.6)

$$\eta = \frac{T_3 - T_4 - T_2 + T_1}{T_3 - T_2}$$

$$= \frac{1273 - 625.9 - 585.8 + 288}{1273 - 585.8} = 0.508 \text{ i.e. } 50.8\%$$

Note – in practice, such a gas turbine engine would have an efficiency nearer to 30%. The difference between this value and the ideal efficiency calculated above can be largely attributed to the effect of friction within the compressor and turbine. This effect is discussed in Chapter 9.

SUMMARY

In this chapter the application of air-standard cycles to power producing engines has been discussed. The key terms that have been introduced are:

air-standard cycle
constant temperature cycle
Carnot steady flow air engine
constant volume cycle

Otto air-standard cycle
compression ratio
detonation
octane rating of petrol
constant pressure cycle
closed-cycle gas turbine engine
pressure ratio
turbine inlet temperature
open-cycle gas turbine engine

Key equations that have been introduced.

For the constant volume cycle,

$$\eta = 1 - \frac{(T_4 - T_1)}{(T_3 - T_2)} \tag{8.1}$$

The compression ratio is defined as

$$r = \frac{v_1}{v_2} = \frac{v_4}{v_3} \tag{8.2}$$

and

$$\eta = 1 - \frac{1}{r^{(\gamma - 1)}} \tag{8.3}$$

For the constant pressure cycle,

$$w_{net} = C_P (T_3 - T_4) - C_P (T_2 - T_1) \tag{8.4}$$

and

$$\eta = \frac{T_3 - T_4 - T_2 + T_1}{T_3 - T_2} \tag{8.6}$$

The optimum pressure ratio is defined by the criterion

$$T_2 = \sqrt{T_3 \times T_1} \tag{8.7}$$

PROBLEMS

Air-standard cycle

1. An engine operates on a Carnot cycle with 0.5 kg of air in a closed system. The temperature limits of the cycle are 800 K and 300 K. If the specific heat input during each cycle is 64 kJ/kg, how many cycles must the engine complete during a second to produce a power output of 40 KW?

2. An engine operates on an air-standard cycle consisting of the following processes:

   1–2, reversible adiabatic compression;
   2–3, heat addition of 870 kJ/kg at constant pressure;
   3–4, reversible adiabatic expansion;
   4–1, heat rejection of 450 kJ/kg at constant volume.

   Sketch the cycle on a P–v diagram. Find the specific work output and thermal efficiency.

3. A spark ignition engine operates on an Otto cycle with a compression ratio of 6. The fuel used is natural gas, with an energy content of 35 000 kJ/m^3. If the engine consumes 2 l/s of gas calculate the power output. Take $\gamma = 1.4$ for air. Note that 1 m^3 = 1000 l.

4. A petrol engine operates on an Otto air-standard cycle with a compression ratio of 8. The engine completes 25 cycles/s and 0.002 kg of air enters at the start of each cycle. If the ambient temperature is 17°C and the maximum temperature in the cycle is 1500 K, estimate the power output from the engine.
   Assume $C_v = 0.72$ kJ/kg K, $\gamma = 1.4$ for air.

5. The 'Lenoir' engine was described in Chapter 1 as having poor performance because the air was not compressed before combustion. The Lenoir engine can be modelled by an air-standard cycle comprising three processes:

   1–2, heat input at constant volume;
   2–3, reversible adiabatic expansion;
   3–1, heat rejection at constant pressure.

   Assuming the engine operates with a temperature at 1 of 17°C and the overall volume ratio is 4, calculate the thermal efficiency.
      Compare with the thermal efficiency of the Otto cycle operating with the same volume ratio. Take $\gamma = 1.4$ for air.

Gas turbine cycle

6. Calculate the thermal efficiency of a closed-cycle gas turbine engine operating between a minimum temperature of 300 K and a maximum temperature of 1100 K. Assume the pressure ratio to be 12 and the working fluid within the system to be carbon dioxide. Take $\gamma = 1.3$ for carbon dioxide.

7. Calculate the thermal efficiency and specific work output for the engine defined in problem 6, if the pressure ratio is changed to the optimum value. Take $C_P = 0.9$ kJ/kg K for carbon dioxide.

8. A stationary gas turbine engine operates on an air-standard open-cycle. Air enters the compressor at 100 kPa and 20°C with a mass flow rate of 12 kg/s. If the maximum pressure and temperature within the cycle are 1.4 MPa and 1250 K respectively, calculate the thermal efficiency and power output. Take $C_P = 1.005$ kJ/kg K and $\gamma = 1.4$ for air.

9. An industrial gas turbine engine takes air into the compressor at 290 K and operates with a pressure ratio of 16 and turbine inlet temperature 1300 K. If the engine has a power output of 5 MW, and operates on an air-standard cycle, calculate:

   (a) the thermal efficiency;
   (b) the fuel consumption if the fuel has an energy content of 44 000 kJ/kg;
   (c) the ratio of the work produced by the turbine compared to the compressor.

   Assume $C_P = 1.05$ kJ/kg K and $\gamma = 1.38$ for air.

Gas turbine engines and propulsion

9

9.1 AIMS

- To compare the thermal efficiency for an industrial gas turbine engine with that for the air-standard cycle.
- To introduce the open-cycle with friction for an industrial gas turbine engine.
- To compare the performance of the cycle with friction with the air-standard cycle.
- To discuss the operation of a two-shaft industrial gas turbine engine.
- To introduce the different types of aircraft gas turbine engines:

  turboprop
  turbojet
  turbofan

- To define the performance parameters for propulsive devices:

  thrust
  propulsive efficiency

- To compare the operating characteristics of aircraft gas turbine engines.
- To introduce the ideal cycle for a turbojet engine.
- To discuss the analysis and application of ram jet engines.

9.2 GAS TURBINE ENGINES

The previous chapter discussed air-standard cycles with particular reference to gas turbine engines. Such engines are widely used for aircraft propulsion and within industry. Aircraft propulsion is covered later in the chapter, so the present discussion will concentrate on industrial gas turbine engines.

9.2.1 Industrial gas turbine engines

Industrial gas turbine engines are used for various applications in the gas, oil and electrical generation industries. The advantage of a gas turbine engine over a reciprocating engine is that it is composed of rotating devices that are well balanced. This means that the gas turbine engine can have a much higher rotational speed than the reciprocating engine and is, therefore, more compact for a given power output. The stationary power-producing gas turbine engine comes into its own for power outputs of between 2 and 25 MW. Below 2 MW the intricate construction of a gas turbine engine generally results in a higher capital cost than a reciprocating engine. Above 25 MW there is an economic case for using a closed-cycle steam plant for power production instead of a gas turbine engine.

Gas turbine engines have the advantage of using a very convenient working fluid, the air from the atmosphere. In addition they will operate on a wide range of fuels. For industrial applications the fuels are, generally, natural gas and liquid fuels, although some experiments have been carried out using finely pulverised coal. However, using coal for internal combustion within an engine has the disadvantage of depositing ash on the turbine blades, with a resultant deterioration in the overall performance. As the balance of fossil fuel reserves changes in the 21st century it may be necessary to develop industrial gas turbine engines burning coal, perhaps with combustion **external** to a heat exchanger instead of inside a combustion chamber.

Looking at industrial gas turbine engines in the context of the discussion in section 8.5, it is necessary to ask how relevant the air-standard cycle is to analysing such engines. Taking the gas turbine engine defined in example 8.8, the maximum temperature and pressure ratio are typical of industrial gas turbine engines. However, the thermal efficiency of 50.8% for the air-standard cycle is much higher than the operational efficiency that would be achieved by such an engine in practice.

Figure 9.1 shows a comparison of typical operational efficiency that can be achieved in an industrial gas turbine engine with that predicted by the air-standard cycle. Taking a pressure ratio of 12 the thermal efficiency predicted by the air-standard cycle is 50.8%, whereas the actual efficiency is likely to be in the region of 30%. This represents a significant difference, which can be attributed to two assumptions used in modelling the air-standard cycle:

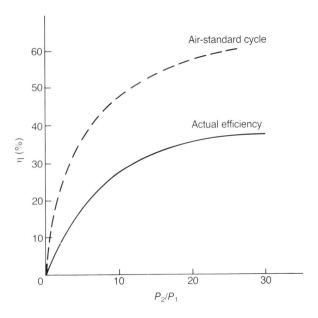

Figure 9.1 Comparison of actual with air-standard efficiency.

1. air is a perfect gas.
2. the components operate with reversible processes.

Looking at assumption 1 first, air is a real gas and although it behaves like a perfect gas over a small temperature range, with the temperatures encountered in a gas turbine engine, the value of C_P will vary quite significantly. The actual variation is shown in Figure 4.10, whereas a perfect gas has a constant value of C_P. This variation will also cause the index of the adiabatic processes γ to change:

C_P (kJ/kg K)	γ (for air)
1.005	1.4
1.1	1.35

The change to both C_P and γ will clearly have an effect on the analysis of a gas turbine cycle and, although it is outside the scope of this text, it does account for part of the difference between the actual and the air-standard efficiency.

However, the major difference between the air-standard cycle and the analysis of an actual engine is due to friction within the compressor and turbine. The air-standard cycle is based upon assumption 2 above and the processes within the compressor and turbine are assumed to undergo reversible adiabatic changes. In reality, both components will operate with some friction during the processes and will have adiabatic efficiencies of less than 1.0.

9.2.2 Gas turbine cycle with friction

The effect of friction in both the compressor and turbine is shown on the open-cycle given in Figure 9.2.

Comparing the cycle with friction with the air-standard cycle, instead of the air leaving the compressor at temperature T_2, it leaves at an actual temperature T_{2a}. Similarly, the air leaves the turbine at temperature T_{4a} instead of T_4.

The adiabatic efficiency of the turbine is defined by equation (6.13):

$$\eta_T = \frac{w_a}{w_{rev}}$$

This relationship states that the actual work out of the turbine will be **less than** if it operated on a reversible adiabatic process. It can be expressed in terms of the enthalpy change across the turbine:

$$\eta_T = \frac{h_3 - h_{4a}}{h_3 - h_4}$$

For a gas turbine this can be written as

$$\eta_T = \frac{C_P(T_3 - T_{4a})}{C_P(T_3 - T_4)}$$

Since air is assumed to be a perfect gas, C_P is constant and the adiabatic efficiency of the turbine is defined by:

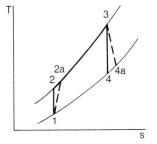

Figure 9.2 T–s diagram for an open-cycle with friction

$$\eta_{\text{T}} = \frac{T_3 - T_{4\text{a}}}{T_3 - T_4} \tag{9.1}$$

Similarly, the adiabatic efficiency of the compressor is defined by equation (6.14):

$$\eta_{\text{c}} = \frac{w_{\text{rev}}}{w_{\text{a}}}$$

This relationship states that the actual work input to the compressor must always be greater than if it operated on a reversible adiabatic process. Like the adiabatic efficiency of the turbine, the adiabatic efficiency of the compressor can be expressed in terms of the temperatures:

$$\eta_{\text{c}} = \frac{T_2 - T_1}{T_{2\text{a}} - T_1} \tag{9.2}$$

The net work output of the cycle with friction is the sum of the work from the turbine and the compressor:

$$w_{\text{net}} = C_{\text{P}}(T_3 - T_{4\text{a}}) - C_{\text{P}}(T_{2\text{a}} - T_1) \tag{9.3}$$

The heat input to the cycle can be evaluated in the same way, so that

$$q_{\text{in}} = C_{\text{P}}(T_3 - T_{2\text{a}}) \tag{9.4}$$

Combining equations (9.3) and (9.4) gives a relationship for the thermal efficiency:

$$\eta = \frac{C_{\text{P}}(T_3 - T_{4\text{a}}) - C_{\text{P}}(T_{2\text{a}} - T_1)}{C_{\text{P}}(T_3 - T_{2\text{a}})}$$

Since air is assumed to be a perfect gas, C_{P} is taken to be constant and the thermal efficiency for the cycle with friction can be expressed as

$$\eta = \frac{T_3 - T_{4\text{a}} - T_{2\text{a}} + T_1}{T_3 - T_{2\text{a}}} \tag{9.5}$$

Example 9.1

An industrial gas turbine engine has an air temperature of 15°C at the compressor inlet and operates with the following conditions:

mass flow rate 10 kg/s;
pressure ratio 12;
turbine inlet temperature 1000°C;
compressor efficiency 100%;
turbine efficiency 88%.

Calculate the power output and thermal efficiency of the engine. Assume air to have the properties $C_{\text{P}} = 1.005$ kJ/kg K and $\gamma = 1.4$.

Conceptual model

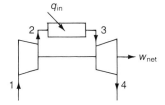

Process diagram — Figure 9.2.

Analysis

$T_1 = 15 + 273 = 288$ K
$T_3 = 1000 + 273 = 1273$ K

For compressor 1–2,

$T_2 = 288 \times (12)^{0.4/1.4} = 585.8$ K

The compressor efficiency is 100% so that

$T_{2a} = T_2 = 585.8$ K

For turbine 3–4,

$$T_4 = 1273 \left(\frac{1}{12}\right)^{0.4/1.4} = 625.9k$$

Applying the turbine efficiency, from equation (9.1),

$$\eta_T = \frac{T_3 - T_{4a}}{T_3 - T_4}$$

$$0.88 = \frac{1273 - T_{4a}}{1273 - 625.9}$$

$$T_{4a} = 1273 - 0.88(1273 - 625.9)$$

$$= 703.6 \text{ K}$$

Substituting in equation (9.3)

$w_{net} = C_P (T_3 - T_{4a}) - C_P (T_{2a} - T_1)$
$w_{net} = 1.005 (1273 - 703.6) - 1.005 (585.8 - 288)$
$\quad\quad = 273$ kJ/kg

Therefore,

power output $= m^{\cdot} \times w_{net}$
$\quad\quad\quad\quad\quad\quad = 10 \times 273 = 2730$ kW i.e. 2.7 MW

Substituting the temperatures in equation (9.5)

$$\eta = \frac{T_3 - T_{4a} - T_{2a} + T_1}{T_3 - T_{2a}}$$

$$= \frac{1273 - 703.6 - 585.8 + 288}{1273 - 585.8} = 0.395 \text{ i.e. } 39.5\%$$

Note — comparing the performance with that for the air-standard cycle, given in example 8.8, including a turbine efficiency of 88% has reduced the power output from 3.5 MW to 2.7 MW and the thermal efficiency from 50.8% to 39.5%.

Example 9.2

Recalculate the power output and thermal efficiency for the industrial gas turbine engine defined in example 9.1, assuming that the compressor efficiency is now 83%.

Conceptual model — as for example 9.1.

Process diagram — Figure 9.2.

Analysis — taking values from example 9.1:

$$T_1 = 288 \text{ K}$$
$$T_2 = 585.8 \text{ K}$$
$$T_3 = 1273 \text{ K}$$
$$T_{4a} = 703.6 \text{ K}$$

Taking a compressor efficiency of 83% and substituting in equation (9.2):

$$\eta_c = \frac{T_2 - T_1}{T_{2a} - T_1}$$

$$0.83 = \frac{585.8 - 288}{T_{2a} - 288}$$

$$T_{2a} = \frac{585.8 - 288}{0.83} + 288 = 646.8 \text{ K}$$

Substituting in equation (9.3):

$$w_{net} = C_p(T_3 - T_{4a}) - C_p(T_{2a} - T_1)$$
$$= 1.005(1273 - 703.6) - 1.005(646.8 - 288)$$
$$= 211.7 \text{ kJ/kg}$$

Therefore,

$$\text{power output} = m \times w_{net}$$
$$= 10 \times 211.7 = 2117 \text{ kW i.e. } 2.1 \text{ MW}$$

Substituting the temperatures in equation (9.5):

$$\eta = \frac{T_3 - T_{4a} - T_{2a} + T_1}{T_3 - T_{2a}}$$

$$= \frac{1273 - 703.6 - 646.8 + 288}{1273 - 646.8} = 0.336 \text{ i.e. } 33.6\%$$

9.2.3 Comparison of the gas turbine engine cycles

The results from examples 8.8, 9.1 and 9.2, which relate to a gas turbine engine with a pressure ratio of 12 and a turbine inlet temperature of 1000°C are shown in Table 9.1.

Before looking at the effect of component efficiency on the overall performance it is necessary to ask how relevant are the values of compressor and turbine efficiency that have been assumed. The answer is that they **are typical** of modern gas turbine engines. Even with the early Whittle jet engines, compressor efficiencies of nearly 80% were both achieved and necessary for the operation of a jet engine. Modern compressors achieve adiabatic efficiency values in the region of 83–85%. Turbine efficiency values are even higher and an uncooled turbine would probably achieve an adiabatic efficiency of 90%. When the turbine is cooled, which is necessary to operate with high turbine inlet temperatures, the cooling air re-enters the main gas stream causing an increase in friction and turbulence. This causes a slight reduction in the adiabatic efficiency.

The first point that emerges from the above comparison is that including the component efficiencies for the compressor and turbine in the cycle analysis, gives a more realistic value of thermal efficiency. The value of 33.6%, resulting from including both the compressor and turbine efficiencies, is still higher than would be achieved in practice. This difference can be attributed to the analysis in example 9.2, being carried out assuming that air is a perfect gas, whereas, in practice it is a real gas. Nevertheless, the cycle with friction provides a more realistic basis for analysing gas turbine engines than the air-standard cycle.

Table 9.1 Gas turbine engine with pressure ratio of 12 and turbine inlet temperature of 1000°C

Example	Compressor efficiency (%)	Turbine efficiency (%)	η	Power (MW)	w_C/w_T
8.8	100	100	50.8	3.5	0.46
9.1	100	88	39.5	2.7	0.52
9.2	84	88	33.6	2.1	0.63

The second point is the marked decrease in the power output when including the component efficiency. With perfect components the power output was 3.5 MW, which drops to 2.1 MW when the compressor and turbine efficiencies are included — a drop of some 40%. This can be attributed to the ratio of the work of the components, as listed in the final column of Table 9.1.

The net work output is the difference between the turbine work output and the compressor work input. In a gas turbine engine the work required by the compressor is a significant proportion of the work output of the turbine, w_c/w_T. As the adiabatic efficiency of the compressor decreases, the work required by the compressor increases. As the adiabatic efficiency of the turbine decreases the work output also decreases. If the efficiency values are sufficiently reduced, a situation is reached where all the work of the turbine is being consumed by the compressor and there is no power output from the engine. In order to maximize both the power output and the thermal efficiency of the engine, the adiabatic efficiency values of the compressor and turbine must be as high as possible.

9.2.4 Twin-shaft gas turbine engines

The type of gas turbine engine discussed so far has a single turbine and a single compressor joined by one single drive shaft. The work from the turbine is greater than that required by the compressor and the difference results in a net work output that is transmitted to the surroundings by the same shaft. This is, appropriately, called a single-shaft gas turbine engine and is shown in Figure 9.3(a).

Where the output from the gas turbine engine is required to operate at constant speed for, say, electrical generation, the single-shaft arrangement is ideal. However, where the engine has to operate over a wide speed range, as for power drives or traction applications the single-shaft arrangement has certain limitations.

With the single-shaft gas turbine engine any variation in the rotational speed will effect both the compressor and turbine. In order to achieve good adiabatic efficiencies these components are designed to operate at one optimum speed. A wide speed range of operation causes the performance, particularly of the compressor, to fall. In order to ensure efficient operation it is preferable to maintain the compressor, and the turbine driving it, at constant speed and to use a separate turbine for the power output. Such a gas turbine engine has a twin-shaft arrangement, as shown in Figure 9.3(b).

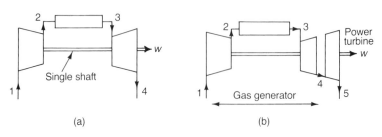

Figure 9.3 Gas turbine engine arrangements.

The turbine, 3–4, drives the compressor and just produces sufficient work to supply the input to the compressor. At the outlet of the turbine at state 4, the gas still has a high energy content as the temperature and pressure will be **well above** atmospheric. This high energy gas stream can be used to drive a separate turbine, 4–5. Since the whole purpose of the compressor-combustion chamber-turbine system between states 1 and 4, is to produce hot gas it is called a 'gas-generator'. The additional turbine after the gas-generator is there to provide the power output and is called a 'power-turbine'.

The analysis of a twin-shaft gas turbine engine is quite complex if the adia-batic efficiencies of the compressor and both turbines are taken into account. However, if the individual components are considered to be ideal, i.e. the adiabatic efficiencies are assumed to be 100%, the analysis of the twin-shaft arrangement is no more complex than the basic air-standard cycle.

Figure 9.4 shows the ideal cycle for a twin-shaft gas turbine engine. Within the gas-generator the work of the turbine just equals the work input to the compressor, so the work can be related using equation (8.4):

$$0 = C_P(T_3 - T_4) - C_P(T_2 - T_1)$$

Since air is assumed to be a perfect gas, the values of C_P are constant and the temperature changes across the components can be related by:

$$T_3 - T_4 = T_2 - T_1 \tag{9.6}$$

This enables the temperature at state 4 to be evaluated. The temperature at state 5 can be evaluated by assuming that the pressure ratio P_2/P_1 is equal to P_3/P_5.

Therefore,

$$\frac{T_3}{T_5} = \frac{T_2}{T_1}$$

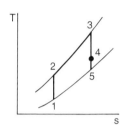

Figure 9.4 T–s diagram for a twin-shaft gas turbine engine.

and

$$T_5 = \frac{T_3 T_1}{T_2} \tag{9.7}$$

In order to evaluate the thermal efficiency of the engine, the net work output for the twin-shaft arrangement is the work from the power turbine:

$$w_{45} = C_P (T_4 - T_5) \tag{9.8}$$

The heat input is given by

$$q_{23} = C_P (T_3 - T_2)$$

and the thermal efficiency is

$$\eta = \frac{C_P (T_4 - T_5)}{C_P (T_3 - T_2)}$$

Since the value of C_P is taken to be constant, this can be simplified to the form

$$\eta = \frac{T_4 - T_5}{T_3 - T_2} \tag{9.9}$$

Example 9.3

A twin-shaft gas turbine engine operates with the following conditions:

air inlet temperature 15°C;
pressure ratio 12;
maximum cycle temperature 1000°C.

Assuming ideal operation of the engine, calculate the thermal efficiency. Take $\gamma = 1.4$ for air.

Conceptual model — Figure 9.3(b).

Process diagram — Figure 9.4.

Analysis

$$T_1 = 15 + 273 = 288 \text{ K}$$
$$T_3 = 1000 + 273 = 1273 \text{ K}$$

For compressor 1–2

$$T_2 = 288 \times (12)^{0.4/1.4} = 585.8 \text{ K}$$

For gas generator turbine 3–4, from equation (9.6),

$$T_3 - T_4 = T_2 - T_1$$
$$1273 - T_4 = 585.8 - 288$$
$$T_4 = 975.2 \text{ K}$$

For power turbine 4–5, from equation (9.7),

$$T_5 = \frac{T_3 T_1}{T_2} = \frac{1273 \times 228}{585.8} = 625.8 \text{ K}$$

Substituting the temperatures in equation (9.9):

$$\eta = \frac{975.2 - 625.8}{1273 - 585.8} = 0.508 \text{ i.e. } 50.8\%$$

Note — this analysis can be compared with that for an ideal single-shaft arrangement, given in example 8.8. Since both cycles are assumed to be ideal, the performance of both the single- and twin-shaft arrangements are the same. In practice, the component efficiencies mean there is a small difference between the two.

9.3 AIRCRAFT GAS TURBINE ENGINES

There are three types of aircraft gas turbine engine — turboprop, turbojet and turbofan. Schematic diagrams of all three types are given in Figure 9.5.

In principle, a turboprop engine is an aerial version of the industrial gas turbine engine. Its purpose is to produce power to drive a propeller. Since the gas turbine engine rotates at a higher speed than that required by the propeller, a single-shaft arrangement requires a reduction gear box between the output shaft

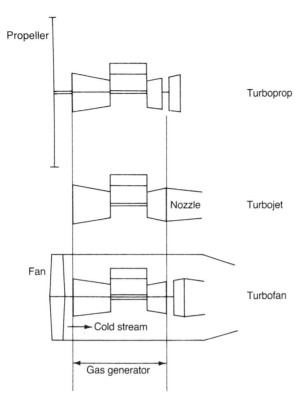

Figure 9.5 Types of aircraft gas turbine engines.

and the propeller. Alternatively, a turboprop engine can operate as a two-shaft arrangement with the power turbine driving the propeller at a lower speed than the gas-generator.

A turbojet engine consists of a gas-generator and a nozzle. The gas-generator produces hot gas which is expanded through a nozzle to produce a high jet velocity. The basic components are shown in Figures 1.7 and 1.8, both of which show turbojet engines. The first shows an engine produced in the 1940s to a Whittle design and incorporates a centrifugal compressor having a low pressure ratio. The latter shows the 'Olympus' turbojet engine, used in Concorde, incorporating a large axial-flow compressor having a high pressure ratio.

A turbofan engine represents a size compromise between the turboprop and the turbojet engines. The heart of the engine is a gas-generator. Gas flows from the gas-generator to a low-pressure turbine driving a fan at the front of the engine. The air from the fan divides, part going through the gas-generator, the rest flowing through a duct bypassing the gas-generator. Since this latter flow of air is unheated by the combustion process it is referred to as the 'cold stream'. The ratio of air in the cold stream compared to that flowing through the gas-generator is defined by the 'bypass ratio'.

To see where all three types of engine fit into the field of aircraft propulsion it is necessary to define the basic performance parameters for propulsive devices.

9.3.1 Propulsion

Propellers, turbojet and turbofan engines all operate on the same basic principle. They take in air at a particular velocity and exhaust the air to atmosphere at a higher velocity, so producing a propulsive force called the 'thrust'.

A propeller can be modelled as shown in Figure 9.6(a). Although a propeller employs several blades, they are assumed to form a complete disc when rotating. Air enters the propeller at V, the forward velocity of the aircraft, and leaves at a higher velocity V_e. The increase in velocity produces the thrust, F. For the following analysis it is assumed that the air on both sides of the propeller is at the same atmospheric pressure. Also, that the air leaves the propeller in an axial direction with no rotation.

It was shown, in section 2.7.3, that the force produced by a steady flow is equal to the rate of change of momentum, $\dot{m}V$. Applying this to the propeller shown in Figure 9.6(a), the air entering the propeller produces a force of magnitude $\dot{m}V$ and acting to the right of the diagram. The air leaving will produce a force of magnitude $\dot{m}V_e$, acting to the left. Since V_e is greater than V, the resultant thrust acts to the left and has magnitude

$$F = \dot{m}V_e - \dot{m}V$$
$$= \dot{m}\,(V_e - V) \tag{9.10}$$

A turbojet, or turbofan, engine can be modelled as shown in Figure 9.6(b). Providing that the expansion in the nozzle is down to atmospheric pressure (a pressure difference would create an additional force, that is ignored in the present discussion), and that the velocity is constant across the outlet, the resultant thrust can also be evaluated using equation (9.10).

Studying equation (9.10) indicates that the thrust is achieved by a mass flow rate of air through the device, associated with an increase in velocity. A particular thrust can, therefore, be achieved by having a large mass flow rate of air and a small increase in velocity. Conversely, the same thrust can be achieved by a small mass flow rate of air and a high increase in velocity. The question is, which is the better way of achieving the thrust? To answer this, it is necessary to introduce another performance parameter, the 'propulsive efficiency'.

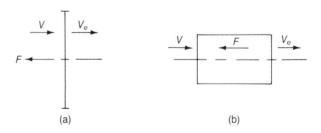

(a)   (b)

Figure 9.6 Propulsive devices.

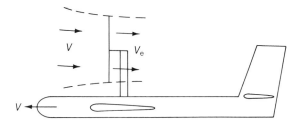

Figure 9.7 Aircraft and propeller.

Consider a propeller as part of an aircraft system, as shown in Figure 9.7. The aircraft is moving forward at velocity, V relative to an observer on the ground. Relative to the propeller, the air is entering at velocity V and leaving at velocity V_e. This gives a resultant thrust F which propels the aircraft at velocity V. The rate of work done on the aircraft is

$$W_a = F \times V$$

and substituting for the thrust F, from equation (9.10), gives

$$W_a = \dot{m} V (V_e - V) \qquad (9.11)$$

To achieve this thrust, the velocity of the air is increased across the propeller. This means that the kinetic energy of the air flow is increased and the power input to the propeller must equal the rate of change of kinetic energy:

$$W_P = \frac{\dot{m}}{2} \left(V_e^2 - V^2 \right) \qquad (9.12)$$

The propulsive efficiency is a measure of the rate of work done on the aircraft compared to the power input to the propeller:

$$\eta_{prop} = \frac{W_a}{W_P}$$

Substituting equations (9.11) and (9.12) gives

$$\eta_{prop} = \frac{\dot{m} V (V_e - V)}{\frac{\dot{m}}{2} (V_e^2 - V^2)}$$

$$= \frac{2V (V_e - V)}{(V_e + V)(V_e - V)}$$

$$= \frac{2V}{V_e + V} \qquad (9.13)$$

This equation for the propulsive efficiency is also true for a turbojet engine. It can also be used for a turbofan engine of the type shown in Figure 9.5, where the cold stream and the gas-generator stream mix before flowing through a single nozzle.

Example 9.4
A propulsive device is required to produce a thrust of 8000 N at a forward velocity of 100 m/s. This can be achieved by either a propeller, with a mass flow rate of 200 kg/s or a turbojet engine with a mass flow rate of 20 kg/s. Compare the propulsive efficiency of these two devices.

Conceptual model — Figure 9.6.

Process diagram — not appropriate.

Analysis — the thrust for both devices is defined by equation (9.10):

$$F = m'(V_e - V)$$

and the propulsive efficiency by equation (9.13):

$$\eta_{prop} = \frac{2V}{V_e + V}$$

For the propeller, applying equation (9.10),

$$8000 = 200 \, (V_e - 100)$$

Therefore,

$$V_e = 140 \text{ m/s}$$

Substituting in equation (9.13)

$$\eta_{prop} = \frac{2 \times 100}{140 + 100} = 0.83 \text{ i.e. } 83\%$$

For the turbojet applying equation (9.10),

$$8000 = 20 \, (V_e - 100)$$

Therefore,

$$V_e = 500 \text{ m/s}$$

Substituting in equation (9.13)

$$\eta_{prop} = \frac{2 \times 100}{500 + 100} = 0.33 \text{ i.e. } 33\%$$

9.3.1. Comparison of propulsive devices

Comparing the propulsive efficiency for the propeller and turbojet considered in example 9.4, it will be seen that there is a considerable difference between the 83% achieved by the propeller and the 33% achieved by the turbojet. Propulsive efficiency is an important criterion, being a measure of how effectively an aircraft engine uses fuel. The higher the propulsive efficiency the lower the fuel consumption. In the above example, the turbojet engine would burn nearly three times the amount of fuel as a turboprop engine, for the same thrust.

The relationship between the mass flow rate and the exhaust velocity is significant. The higher the mass flow rate the lower the velocity increase, V_e–V, and the better the propulsive efficiency. A propeller achieves the high mass flow rate through its much larger diameter compared to the turbojet. If a propeller is so efficient the obvious question is why are there so few propeller-driven aircraft around? To provide an answer it is necessary to look at the operation of the propeller in greater detail.

Figure 9.8 shows the velocity of the air relative to a propeller blade. The incoming air enters in an axial direction with velocity V. However, the blade is itself moving and has a rotational velocity so that, to an observer on the blade, the relative velocity of the air is much higher than V. If the forward velocity is, say, 200 m/s then the velocity of the air flowing over the blade might have a velocity of 300 m/s. At this sort of velocity the air will be very close to the local speed of sound, resulting in shock waves developing on the blade. Shock waves cause a breakdown in the propulsive efficiency.

Figure 9.9 shows a typical variation of the propulsive efficiency of a propeller. Above a forward velocity of about 200 m/s the efficiency falls off quite rapidly. By comparison, the propulsive efficiency of a turbojet increases with velocity and reaches acceptable operating values above 450 m/s. This is why Concorde is propelled by turbojet engines.

At a forward velocity of between 200 and 450 m/s, there is a region where the propulsive efficiency of both the propeller and turbojet are unacceptably low. It is within this region that turbofan engines are employed. A turbofan engine has a greater mass flow rate than the turbojet engine, requires a lower exhaust velocity and has a higher propulsive efficiency. Also, engine noise is a function of exhaust velocity, the lower exhaust velocities of turbofan engines help reduce airport noise pollution. The turbofan engine can operate at forward velocities near, or above, the speed of sound because the air intake acts as a diffuser and reduces the **actual** velocity of the air entering the fan.

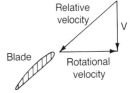

Figure 9.8 Propeller blade velocities.

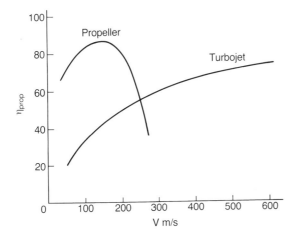

Figure 9.9 Comparison between a propeller and a turbojet.

9.4 TURBOJET ENGINES

Although turbojet engines are still employed for subsonic aircraft (i.e. below the speed of sound), they are generally being superseded by turbofan engines due to their improved propulsive efficiency. Nevertheless, it is proposed to consider the analysis of turbojet engines in preference to turbofan engines. This is for the simple reason that the turbojet cycle is much easier to both comprehend and analyse than the turbofan cycle. As such, the turbojet forms a useful basis for the analysis of all aircraft gas turbine engines. In addition, the turbojet engine was the first of the aircraft gas turbine engines to be developed and played a significant role in the application of jet engines to aircraft propulsion.

The schematic diagram of the turbojet engine, shown in Figure 9.9, is somewhat simplistic. It takes no account of the fact that air may be entering the engine at a wide range of different velocities.

When the aircraft is stationary on the ground the effective forward velocity is zero and the velocity entering the compressor can then be assumed to be zero. Clearly this is impractical because a zero velocity means that there is **no mass flow rate**. Nevertheless, the analysis of a turbojet engine under static conditions is based on the assumption that the velocity through the gas-generator is very low, i.e. $V \approx 0$.

At the other end of the velocity scale an engine in Concorde operating at twice the speed of sound will have a forward velocity of around 600 m/s. It is impossible to design engine components to operate efficiently if they are required to cope with such a wide range of velocities. Some device is required to control the velocity of the air entering the compressor. Such a device is a diffuser. In practice, the air intake for the engine acts as a diffuser and reduces the velocity of the air entering the compressor to within acceptable limits.

Figure 9.9 shows a schematic diagram of a turbojet engine. As such, it is an improvement over that given in Figure 9.5, as it now incorporates a diffuser. Stations 1–4 represent the gas-generator. Ahead of the gas-generator is a diffuser, 0–1, to control the velocity of the air entering the compressor. Finally, the air flows through the nozzle, 4–5, to achieve the required exhaust velocity. An ideal turbojet cycle is shown in Figure 9.11.

Within the diffuser the air is reduced from velocity V_0 to V_1 where, for the purposes of analysis, V_1 is taken to be zero. This reduction in velocity brings about an increase in enthalpy of the air flow, shown in Figure 9.11 as an increase in temperature between states 0 and 1. Applying the steady flow energy equation to the diffuser gives equation (5.7):

$$V_0^2 - V_1^2 = 2(h_1 - h_0)$$

But V_1 is assumed to be zero and air is taken to be a perfect gas, so that

$$V_0^2 = 2\,C_P(T_1 - T_0) \tag{9.14}$$

Within the gas-generator the work output of the turbine just balances the work input, so from equation (9.6),

$$T_3 - T_4 = T_2 - T_1$$

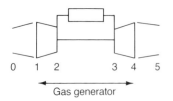

Figure 9.10 Schematic diagram of a turbojet engine.

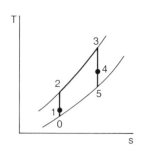

Figure 9.11 Ideal turbojet cycle.

Finally, the air flows through the nozzle where the velocity is increased from V_4 to V_5. Applying equation (5.7)

$$V_5^2 - V_4^2 = 2(h_4 - h_5)$$

Since the velocity through the gas-generator is assumed to be constant, $V_4 = V_1 = 0$. Therefore,

$$V_5^2 = 2\,C_P(T_4 - T_5) \qquad (9.15)$$

Example 9.5

A turbojet engine operates with the following conditions:

air inlet temperature 20°C;
mass flow rate 16 kg/s;
pressure ratio 10;
turbine inlet temperature 1100 K.

Assuming ideal operation of the engine, calculate the static thrust. Assume $\gamma = 1.4$ and $C_P = 1.005$ kJ/kg K for air.

Conceptual model — Figure 9.10.

Process diagram — under static conditions $V_0 = 0$ and $T_0 = T_1$. The cycle is, therefore,

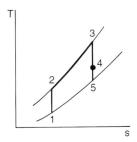

Analysis

$$T_1 = 20 + 273 = 293 \text{ K}$$

For compressor 1–2,

$$T_2 = 293 \times (10)^{0.4/1.4} = 565.7 \text{ K}$$

For gas-generator turbine 3–4, from equation (9.6),

$$T_3 - T_4 = T_2 - T_1$$
$$1100 - T_4 = 565.7 - 293$$
$$T_4 = 827.3 \text{ K}$$

For nozzle 4–5, from equation (9.7),

$$T_5 = \frac{T_3 T_1}{T_2} = \frac{1100 \times 293}{565.7} = 569.7 \text{ K}$$

Applying equation (9.15)

$$V_5^2 = 2\,C_P(T_4 - T_5)$$
$$= 2 \times 1005\,(827.3 - 569.7)$$
$$V_5 = 719.5\ \text{m/s}$$

The thrust is given by equation (9.10):

$$F = m^.(V_5 - V_1)$$
$$= 16\,(719.5 - 0) = 11\,512\ \text{N} \quad \text{i.e. } 11.5\ \text{kN}$$

Example 9.6

Calculate the thrust and propulsive efficiency for the turbojet engine, defined in example 9.5, when operating at a flight velocity of 200 m/s.

Conceptual model — Figure 9.10.

Process diagram — Figure 9.11.

Analysis — for diffuser 0–1, $T_0 = 293$ K and from equation (9.14),

$$V_0^2 = 2\,C_P(T_1 - T_0)$$
$$(200)^2 = 2 \times 1005\,(T_1 - 293)$$
$$T_1 = 312.9\ \text{K}$$

For compressor 1–2,

$$T_2 = 312.9 \times (10)^{0.4/1.4} = 604.1\ \text{K}$$

For gas-generator turbine 3–4, from equation (9.6),

$$T_3 - T_4 = T_2 - T_1$$
$$1100 - T_4 = 604.1 - 312.9$$
$$T_4 = 808.8\ \text{K}$$

For nozzle 4–5,

$$\frac{P_2}{P_0} = \frac{P_3}{P_5}$$

so that

$$\frac{T_2}{T_0} = \frac{T_3}{T_5}$$

and

$$T_5 = \frac{T_3 T_0}{T_2} = \frac{1100 \times 293}{604.1} = 533.5\ \text{K}$$

Applying equation (9.15),

$$V_5^2 = 2\,C_P(T_4 - T_5)$$
$$= 2 \times 1005\,(808.8 - 533.5)$$
$$V_5 = 743.8\ \text{m/s}$$

The thrust is given by equation (9.10):

$$F = m'(V_5 - V_0)$$
$$= 16\,(743.8 - 200) = 8702\ \text{N i.e. } 8.7\ \text{kN}$$

The propulsive efficiency is given by equation (9.13):

$$\eta_{\text{prop}} = \frac{2V_0}{V_5 + V_0}$$

$$= \frac{2 \times 200}{743.8 + 200} = 0.424\ \text{i.e. } 42.4\%$$

Note — comparing the thrust calculated in this example with that found in example 9.5, the thrust in flight is less than the static thrust and this is true for all turbojet engines (and propellers).

9.5   RAM JET ENGINES

From example 9.6 it was shown that reducing the air velocity from 200 m/s down to zero increased the air temperature by 19.9°C. Since this change takes place in a diffuser it can be assumed that the process is adiabatic, i.e. that the heat transfer to the surroundings is negligible. Therefore, the increase in temperature must be accompanied by an increase in pressure.

Applying equation (4.25)

$$\frac{P_1}{P_0} = \left(\frac{T_1}{T_0}\right)^{\gamma/(\gamma-1)}$$

and using data from example 9.6 gives a pressure ratio of

$$\frac{P_1}{P_0} = \left(\frac{312.9}{293}\right)^{1.4/0.4} = 1.26$$

However, the temperature ratio and, therefore, the pressure ratio, increase with the air velocity. This is apparent from a study of equation (9.14), the temperature difference being proportional to the square of the velocity. The higher the velocity the greater the temperature difference.

Figure 9.12 shows the variation of pressure ratio achieved in the diffuser at different air velocities. The graph is based on the ideal process assumed in example 9.6 and an air temperature of 293 K. Of course, in a real air intake the velocity leaving the diffuser would be greater than zero and the actual pressure ratio would be less than that given in Figure 9.12. Nevertheless, the curve demonstrates very clearly how the pressure ratio increases with velocity. At a velocity of 600 m/s, which is roughly the cruising speed of Concorde, the ideal pressure ratio achieved in the air intake is about 5. This is in addition to the pressure ratio in the compressor of the turbojet engine.

At higher velocities the pressure ratio is even greater. For example, at an air velocity of 750 m/s the ideal pressure ratio exceeds 10. At this pressure ratio it

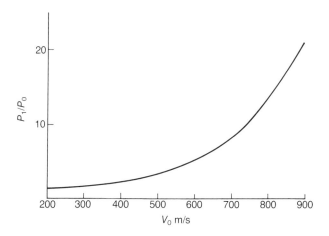

Figure 9.12 Variation of pressure ratio with velocity.

is pertinent to ask whether a compressor is required at all. If the compressor can be dispensed with, this eliminates the need for a turbine. A jet engine without compressor or turbine, achieving the required pressure ratio entirely by the 'ram' effect in intake, is called a ram jet engine and is shown in Figure 9.13.

Air enters the diffuser in which the pressure is increased from 0 to 1, it then enters the combustion chamber and is heated between 1 and 2, finally expanding in the nozzle to state 3. The ideal ram jet engine cycle is shown in Figure 9.13(b) and is straightforward to analyse using equations (9.13) and (9.14), as illustrated in example 9.7 below.

The ram jet engine is a very effective way of achieving thrust at high forward velocities. Unfortunately, the aircraft being propelled has to achieve the high forward velocities before the ram jet engine can come into operation. This is its main disadvantage, the need of ancillary engines to accelerate the aircraft up to a high enough speed at which it is possible to switch over to the ram jet engine. One solution that has been suggested is a 'hybrid' engine, combining the characteristics of a turbojet at low velocities and a ram jet at

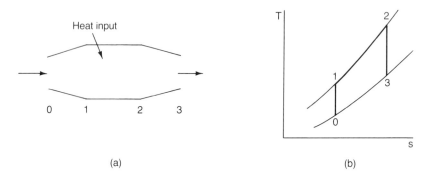

Figure 9.13 Ram jet engine.

high velocities. As the velocity increases more of the air bypasses the gas-generator to flow through a separate combustion chamber before expanding through the nozzle. If a supersonic airliner is to be built with speeds well in excess of those achieved by Concorde, then it would operate with such a hybrid engine.

Example 9.7

An experimental aircraft is to be propelled by a ram jet engine at a flight velocity of 1000 m/s. The engine is to operate under ideal conditions with:

air inlet temperature 250 K;
mass flow rate 20 kg/s;
maximum cycle temperature 1500 K.

Calculate the thrust and propulsive efficiency. Take $C_P = 1.005$ kJ/kg K.

Conceptual model — Figure 9.13(a).

Process diagram — Figure 9.13(b).

Analysis — For diffuser 0–1, from equation (9.14),

$$
\begin{aligned}
V_0^2 &= 2\,C_P(T_1 - T_0) \\
(1000)^2 &= 2 \times 1005\,(T_1 - 250) \\
T_1 &= 747.5 \text{ K}
\end{aligned}
$$

For nozzle 2–3,

$$
\frac{P_1}{P_0} = \frac{P_2}{P_3}
$$

so that

$$
\frac{T_1}{T_0} = \frac{T_2}{T_3}
$$

and

$$
T_3 = \frac{T_2 T_0}{T_1} = \frac{1500 \times 250}{747.5} = 501.7 \text{ K}
$$

Applying equation (9.15),

$$
\begin{aligned}
V_3^2 &= 2\,C_P(T_2 - T_3) \\
&= 2 \times 1005\,(1500 - 501.7) \\
&= 1416.5 \text{ m/s}
\end{aligned}
$$

The thrust is given by equation (9.10):

$$
\begin{aligned}
F &= m(V_3 - V_0) \\
&= 20\,(1416.5 - 1000) = 8330 \text{ N i.e. } 8.3 \text{ kN}
\end{aligned}
$$

The propulsive efficiency is given by equation (9.13):

$$
\eta_{\text{prop}} = \frac{2 \times 1000}{1416.5 + 1000} = 0.83 \text{ i.e. } 83\%
$$

SUMMARY

In this chapter the application of gas turbine engines to both power production and aircraft propulsion has been discussed. The key terms that have been introduced are:

industrial gas turbine engine
gas turbine cycle with friction
single-shaft gas turbine engine
twin-shaft gas turbine engine
gas-generator
power-turbine
turboprop engine
turbojet engine
turbofan engine
bypass ratio
propeller
thrust
propulsive efficiency
turbojet engine cycle
ram jet engine

Key equations that have been introduced.

For the gas turbine cycle with friction.
adiabatic efficiency of the turbine:

$$\eta_T = \frac{T_3 - T_{4a}}{T_3 - T_4} \tag{9.1}$$

adiabatic efficiency of the compressor:

$$\eta_c = \frac{T_2 - T_1}{T_{2a} - T_1} \tag{9.2}$$

net work output:

$$w_{net} = C_P(T_3 - T_{4a}) - C_P(T_{2a} - T_1) \tag{9.3}$$

thermal efficiency:

$$\eta = \frac{T_3 - T_{4a} - T_{2a} + T_1}{T_3 - T_{2a}} \tag{9.5}$$

For the ideal twin-shaft gas turbine engine cycle.
gas-generator turbine output equals compressor work input:

$$T_3 - T_4 = T_2 - T_1 \tag{9.6}$$

net work output:

$$w_{net} = C_P(T_4 - T_5) \tag{9.8}$$

thermal efficiency:

$$\eta = \frac{T_4 - T_5}{T_3 - T_2} \tag{9.9}$$

For propulsive devices.
  thrust:

$$F = m'(V_e - V) \tag{9.10}$$

  propulsive efficiency:

$$\eta_{prop} = \frac{2V}{V_e + V} \tag{9.13}$$

For the ideal turbojet engine cycle:
  for the change of temperature in the diffuser:

$$V_0^2 = 2\,C_P(T_1 - T_0) \tag{9.14}$$

  for the increase in velocity in the nozzle:

$$V_5^2 = 2\,C_P(T_4 - T_5) \tag{9.15}$$

PROBLEMS

Industrial gas turbine engines

1. A single-shaft industrial gas turbine engine operates with 10 kg/s of air entering the compressor at 100 kPa and 20°C. If the maximum pressure and temperature in the engine are 1.2 MPa and 1150 K respectively, calculate the thermal efficiency and power output of the engine. Assume a compressor efficiency of 1.0 and a turbine efficiency of 0.87. Take the properties of air to be
   $C_P = 1.005$ kJ/kg K and $\gamma = 1.4$.

2. Recalculate the power output and thermal efficiency for the industrial gas turbine engine defined in problem 1, if the compressor efficiency is 82%.

3. A single-shaft industrial gas turbine engine takes in air at 290 K and operates with the following conditions:

   pressure ratio 15;
   turbine inlet temperature 1250 K;
   compressor efficiency 0.84;
   turbine efficiency 0.89.

   If the engine has a power output of 6 MW, calculate the fuel consumption in kg/s if the fuel has an energy content of 44 000 kJ/kg. Assume $\gamma = 1.38$ for air.

4. A twin-shaft gas turbine engine operates on an ideal cycle with the following conditions:

air inlet temperature 15°C;
air mass flow rate 25 kg/s;
pressure ratio 18;
maximum cycle temperature 1100°C.

Calculate the power output and thermal efficiency of the engine. Assume $C_P = 1.06$ kJ/kg K and $\gamma = 1.37$ for air.

Propulsion

5. A turboprop engine is used to drive a propeller. The propeller is required to provide a thrust of 5000 N at a flight velocity of 150 m/s, with propulsive efficiency of 82%. If the turboprop engine operates with a thermal efficiency of 40%, calculate:

   (a) the power output of the engine;
   (b) the fuel consumption if the fuel has an energy content of 43 000 kJ/kg.

6. A turboprop engine has a twin-shaft arrangement with the power turbine driving the propeller. During take off the propeller has a thrust of 20 000 N at a forward velocity of 30 m/s. Under these conditions the propulsive efficiency is 50%. If the air temperature is 15°C and the engine operates with a pressure ratio of 14 and maximum cycle temperature 1100 K, calculate the mass flow rate of air through the engine.

   Ignore the forward velocity of the aircraft in the analysis of the cycle. Assume $C_P = 1.008$ kJ/kg K and $\gamma = 1.4$ for air.

7. A turbojet engine is being tested under the static conditions on the ground. Calculate the static thrust if the engine operates with the following conditions:

   air inlet temperature 20°C;
   mass flow rate 18 kg/s;
   pressure ratio 14;
   turbine inlet temperature 1300 K.

   Assume the engine operates on an ideal cycle with the following properties for air — $C_P = 1.05$ kJ/kg K and $\gamma = 1.38$.

8. Calculate the thrust and propulsive efficiency for the turbojet engine defined in Problem 6, when operating at a flight velocity of 251 m/s at an altitude where the air temperature is –10°C.

9. A ram jet engine is designed to operate at a flight velocity of 850 m/s. The engine operates under ideal conditions with:

   air inlet temperature 230 K;
   mass flow rate 15 kg/s;
   maximum cycle temperature 1450 K.

   Calculate the thrust and propulsive efficiency. Take $C_P = 1.04$ kJ/kg K and $\gamma = 1.38$ for air.

Mixture of gases 10

10.1 AIMS

- To introduce the concept of a mixture of perfect gases.
- To redefine the equation of state for a perfect gas — as applicable to constituent gases.
- To define Dalton's law of partial pressures.
- To evaluate the thermodynamic properties of a mixture of gases using the Gibbs–Dalton law.
- To introduce psychrometry as the study of mixtures of dry air and water vapour.
- To define the amount of water vapour present through:

  specific humidity
  relative humidity

- To introduce the psychrometric chart as a process diagram.
- To discuss the use of dry-bulb and wet-bulb temperatures for the evaluation of relative humidity.
- To define the dew-point of a mixture of air and water vapour.
- To describe the basic psychrometric processes of heating, dehumidification and humidification.
- To discuss the psychrometric operation of air-conditioners, tumble dryers and cooling towers.

10.2 MIXTURES OF PERFECT GASES

The previous two chapters have been concerned with engines using air as the working fluid. In order to analyse engine performance, air has been considered to be a perfect gas. In reality, air is not a single gas but a mixture of gases. It is made up of roughly 75% nitrogen, 23% oxygen (by mass) with the remainder being small proportions of argon, carbon dioxide, ozone, etc.

Although carbon dioxide and ozone represent just a tiny proportion of air, nevertheless the quantity of these gases in the atmosphere has a significant effect on the environment. The amount of carbon dioxide determines the radiant properties of the air and controls the amount of solar energy trapped within the atmosphere, the so-called 'greenhouse' effect. The amount of ozone determines the size of the 'ozone layer' that filters out harmful ultraviolet rays.

Nevertheless, from a thermodynamic point of view, the minor constituents have a negligible effect on the properties of air. The thermodynamic properties of air can be considered to be solely determined by the major constituents of nitrogen and oxygen. Both of these can be assumed to behave as a perfect gas and, therefore, the resulting mixture also behaves as a perfect gas.

Air is so widely used in thermodynamic situations that its properties are well defined. Nevertheless, air is just one mixture of gases within a wide range of possible mixtures. It would be impractical to define the properties for every conceivable mixture of gases that can be created. Therefore, it is necessary to develop suitable relationships that will define the properties of any mixture, provided that the composition of the mixture and the properties of the individual constituent gases can be defined.

In particular, a large part of this chapter will be devoted to the discussion of mixtures involving air and water vapour, a study of psychrometry. This is necessary for such devices as air-conditioners, tumble dryers and cooling towers.

In order to analyse any mixture of gases it is assumed that each constituent behaves as a perfect gas and that the resulting mixture also behaves as a perfect gas. A perfect gas was defined in Chapter 4 as one that obeys the equation of state

$$Pv = RT$$

where R is the gas constant.

If the value of the gas constant for a particular constituent gas is not known, it can be found using the universal gas constant R_0 from equation (4.8)

$$R = \frac{R_0}{M}$$

where M is the molecular weight of the gas. Substituting the value of the universal gas constant, 8.314 kJ/kmol K,

$$R = \frac{8.314}{M} \tag{10.1}$$

The equation of state for a perfect gas, given above, is valid when considering individual gases. In a mixture of gases, the mass of each constituent gas can be different and this difference in mass must be taken into account.

Multiplying each side of the equation of state by the mass m gives

$$P(mv) = mRT$$

where mv is the actual volume of the gas, given the symbol $\mathscr{V}$. Substituting back into the equation of state,

$$P(\mathscr{V}) = mRT \tag{10.2}$$

Checking that the units are consistent gives

$$\frac{N}{m^2} m^3 = kg \frac{J}{kgK} K$$

with units of J, i.e. Nm, on both sides of the equation.

10.3 PARTIAL PRESSURES

Within a mixture of gases, the total mixture will have a pressure but that pressure will be made up of individual 'part' pressures from the individual constituents. The pressure contributed by each constituent is known as its 'partial pressure' and the total pressure is the sum of the partial pressures of each constituent. This is defined by Dalton's law of partial pressures and it can be visualized using Figure 10.1.

Figure 10.1 shows three equally sized containers. The two on the right-hand side contain two gases, 1 and 2 respectively. Assuming that both are at the same temperature, the pressure of each gas will be P_1 and P_2 respectively. Now, if the two gases are brought together to form a mixture in the left-hand container and the temperature remains the same, each constituent gas will retain its pressure but the mixture will now have a total pressure equal to $P_1 + P_2$.

Clearly, this can only be true if the mixture and the constituent gases have the **same temperature and volume**. This provides a formal definition of Dalton's law as

'The pressure of a mixture of gases is equal to the sum of the partial pressures of its constituent gases, providing that each occupies a volume equal to that of the mixture and is at the same temperature as the mixture'.

Although the present discussion is only concerned with partial pressures, it should be noted that it is also possible to have partial volumes but **not** partial temperatures. Temperatures are incorporated in values of enthalpy or internal energy for each constituent, as discussed later in section 10.4.

Considering a mixture of gases with more than two constituents, the properties of the mixture can be expressed as

mixture = gas 1 + gas 2 + ...

In terms of the mass of each constituent it can be seen that the mass of the mixture is equal to the sum of the individual masses

$$m = m_1 + m_2 + ... \tag{10.3}$$

From Dalton's law, the pressure of the mixture is equal to the partial pressures

$$P = P_1 + P_2 + ... \tag{10.4}$$

Using the form of the equation of state given in equation (10.2),

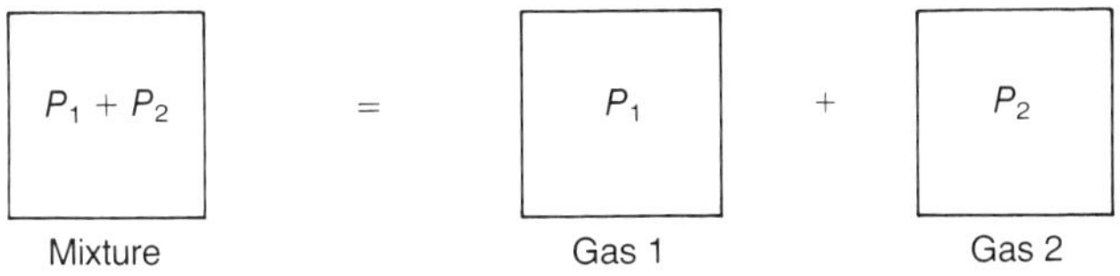

Figure 10.1 A mixture of two gases.

$$P = \frac{mRT}{\mathscr{V}}$$

$$P_1 = \frac{m_1 R_1 T}{\mathscr{V}}$$

$$P_2 = \frac{m_2 R_2 T}{\mathscr{V}}$$

Substituting in equation (10.4),

$$\frac{mRT}{\mathscr{V}} = \frac{m_1 R_1 T}{\mathscr{V}} + \frac{m_2 R_2 T}{\mathscr{V}} + \ldots$$

Since the temperature and volume are the same, it follows that

$$mR = m_1 R_1 + m_2 R_2 + \ldots \tag{10.5}$$

Providing that the gas constant for each component can be defined, equation (10.5) provides a method of finding the gas constant for the whole mixture.

It should also be noted that the mR term for each constituent will be in proportion to the partial pressure for that constituent gas. This is illustrated in the example given below.

Example 10.1

Air can be assumed to consist of a simplified mixture of nitrogen (N_2) and oxygen (O_2) in the proportion 76.7% N_2 to 23.3% O_2, by mass.

Calculate the gas constant for the mixture and the partial pressure for each component if the air pressure is 100 kPa.

Analysis — from equation (10.5)

$$R = \frac{m_1}{m} R_1 + \frac{m_2}{m} R_2$$

Taking N_2 as gas 1, from equation (10.1),

$$R = \frac{8.314}{28} = 0.297 \text{ kJ/kg K}$$

Taking O_2 as gas 2,

$$R = \frac{8.314}{32} = 0.260 \text{ kJ/kg K}$$

Therefore,

$$R = (0.767 \times 0.297) + (0.233 \times 0.26)$$
$$= 0.288 \text{ kJ/kg K}$$

The partial pressures are given by equation (10.4):

$$P = P_1 + P_2$$

which can be expressed as

$$mR = m_1 R_1 + m_2 R_2$$

Combining, gives

$$\frac{P_1}{P} = \frac{m_1 R_1}{mR} = 0.767 \times \frac{0.297}{0.288} = 0.79$$

and

$$\frac{P_2}{P} = \frac{m_2 R_2}{mR} = 0.233 \times \frac{0.26}{0.288} = 0.21$$

i.e. the partial pressure of N_2 is 79 kPa and the partial pressure of O_2 is 21 kPa.

Note — the simplified analysis of air used in this example gives a value of 0.288 kJ/kg K for the gas constant, whereas the true value is 0.287 kJ/kg K.

10.4 PROPERTIES OF GAS MIXTURES

The discussion of partial pressures and, from it, the derivation of equation (10.5) for the gas constant provides the foundation for analysing mixtures of gases. However, on its own, the gas constant is of limited value. Thermodynamic situations require the evaluation of other properties such as the specific enthalpy h and the specific internal energy u.

An extension to Dalton's law of partial pressures, known as the Gibbs–Dalton law, states that such properties of a mixture can be evaluated from the sum of the properties of the constituent gases. Therefore, the enthalpy of a mixture is equal to the enthalpies of the individual constituents:

$$mh = m_1 h_1 + m_2 h_2 + \ldots \tag{10.6}$$

Similarly, the internal energy of the mixture is equal to the internal energies of the individual constituents:

$$mu = m_1 u_1 + m_2 u_2 + \ldots \tag{10.7}$$

It can be shown, using equations (4.12) and (4.13), that the enthalpy and internal energy can be expressed in terms of the specific heats for a perfect gas, so that

$$h = C_p T \quad \text{and} \quad u = C_v T$$

which, when substituted in equations (10.6) and (10.7), gives

$$mC_P = m_1 C_{P_1} + m_2 C_{P_2} + \ldots \tag{10.8}$$

and

$$mC_v = m_1 C_{v_1} + m_2 C_{v_2} + \ldots \tag{10.9}$$

Example 10.2

A rigid container holds 1 kg of a mixture of gas consisting of 40% hydrogen (H_2) and 60% carbon dioxide (CO_2) by mass. If the gas temperature increases from 20°C to 75°C, calculate the heat transferred during this process. Take

$$C_P (H_2) = 14.4 \text{ kJ/kg K}$$
$$C_P (CO_2) = 0.87 \text{ kJ/kg K}$$

Analysis — for a constant volume container, from equation (4.9),

$$q = C_V (T_2 - T_1)$$

From equation (4.16)

$$C_V = C_P - R$$

To evaluate C_V it is necessary to evaluate C_P and R for the mixture. Taking H_2 as gas 1, from equation (10.2),

$$R = \frac{8.314}{2} = 4.157 \text{ kJ/kg K}$$

Taking CO_2 as gas 2, from equation (10.2),

$$R = \frac{8.314}{44} = 0.189 \text{ kJkg K}$$

Substituting in equation (10.5)

$$R = (0.4 \times 4.157) + (0.6 \times 0.189) = 1.776 \text{ kJ/kg K}$$

From equation (10.8)

$$C_P = (0.4 \times 14.4) + (0.6 \times 0.87) = 6.282 \text{ kJ/kg K}$$

Therefore,

$$C_V = C_P - R$$
$$= 6.282 - 1.776$$
$$= 4.506 \text{ kJ/kg K}$$

Substituting in equation (4.9)

$$q = C_V (T_2 - T_1)$$
$$= 4.506 (75 - 20)$$
$$= 247.8 \text{ kJ/kg}$$

Example 10.3

A mixture of gases, consisting of 70% nitrogen (N_2), 10% oxygen (O_2) and 20% carbon dioxide (CO_2) enters a compressor at 50°C and 100 kPa, and is compressed by a reversible adiabatic process to 600 kPa. Find the specific work input assuming

C_P (N_2) = 1.04 kJ/kg K
C_P (O_2) = 0.94 kJ/kg K
C_P (CO_2) = 0.93 kJ/kg K

Conceptual model

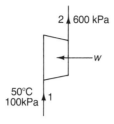

Process diagram

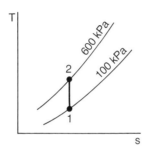

Analysis — from equation (5.9)

$$w = h_1 - h_2$$
$$= C_P (T_1 - T_2)$$

where

$$T_2 = T_1 \left(\frac{P_2}{P_1} \right)^{(\gamma - 1)/\gamma}$$

To find γ it is necessary to evaluate C_P and R, since

$$\gamma = \frac{C_P}{C_v} = \frac{C_P}{C_P - R}$$

Taking N_2 as gas 1,

$$R_1 = \frac{8.314}{28} = 0.297 \text{ kJ/kg K}$$

Taking O_2 as gas 2,

$$R_2 = \frac{8.314}{32} = 0.260 \text{ kJ/kg K}$$

Taking CO_2 as gas 3,

$$R_3 = \frac{8.314}{44} = 0.189 \text{ kJ/kg K}$$

Therefore,

$$R = (0.7 \times 0.297) + (0.1 \times 0.26) + (0.2 \times 0.189)$$
$$= 0.272 \text{ kJ/kg K}$$

From equation (10.8)

$$C_P = (0.7 \times 1.04) + (0.1 \times 0.94) + (0.2 \times 0.93)$$
$$= 1.008 \text{ kJ/kg K}$$

Therefore,

$$\gamma = \frac{1.008}{1.008 - 0.272} = 1.37$$

and

$$T_2 = T_1 \left(\frac{P_2}{P_1} \right)^{(\gamma - 1)/\gamma}$$

$$= 323 \left(\frac{600}{100} \right)^{0.37/1.37} = 524 \text{ K}$$

The specific work input is

$$w = C_P (T_1 - T_2)$$
$$= 1.008 (323 - 524)$$
$$= 202.6 \text{ kJ/kg}$$

10.5 PSYCHROMETRY

Air has been described as a mixture of gases in which nitrogen and oxygen are the main constituents. Its composition can be considered as fixed. The amount of each gas in the mixture is a fixed percentage, determined either on the basis of mass or volume.

In reality, air also contains water vapour. This is evident from the creation of clouds or the depositing of dew on the ground during the night. The amount of water vapour is not fixed, but varies according to the different weather conditions. The amount of water vapour in the air at any given time is determined by its 'humidity'. The study of mixtures of air and water vapour is termed 'psychrometry'.

10.5.1 Humidity

Before giving a precise definition of humidity, it is necessary to consider how a mixture of air and water vapour can be analysed. If the air and water vapour are considered to be separate constituents of a mixture, then the air can be referred to as 'dry' air. Although dry air is itself a mixture of gases it can, for the purpose of this discussion, be considered as a single homogeneous perfect gas with well defined properties. Therefore, dry air can be taken to be one of the constituents of an air and water vapour mixture.

The other constituent, water vapour, can be treated as a separate constituent of the total mixture. Under normal conditions, the water vapour in the atmosphere cannot be seen. This is because it is in a superheated state. As an example, consider a condition in which the partial pressure of the water vapour within the mixture is, say, 1.7 kPa. At this pressure the water vapour alone would have a saturation temperature of 15°C. If the ambient temperature is above 15°C the water vapour must be within the superheat region. To take the example a stage further, if the ambient temperature is 20°C, the water vapour will have 5 K of superheat. This situation is illustrated in Figure 10.2.

Within the superheat region, the water vapour can be assumed to behave as a perfect gas. In practice, this is not strictly true, but the assumption is particular useful as it allows the mixture of air and water vapour to be analysed as a mixture of two perfect gases. On the basis of this assumption it is now possible to define the humidity of the mixture. There are two types of humidity that are used with relation to the composition of the mixture, 'specific humidity' and 'relative humidity'.

Specific humidity is defined as 'the mass of water vapour present in each unit mass of dry air of the mixture' and is given the symbol ω. Using the subscripts 'v' and 'a' to denote the water vapour and dry air respectively:

$$w = \frac{m_v}{m_a} \tag{10.10}$$

From the equation of state for a perfect gas, given in equation (10.2),

$$m_v = \frac{P_v \mathscr{V}}{R_v T}$$

and

$$m_a = \frac{P_a \mathscr{V}}{R_a T}$$

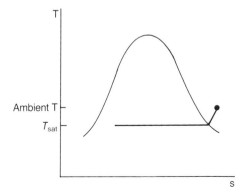

Figure 10.2 Superheated water vapour.

Substituting in equation (10.10) and cancelling out $\mathscr{V}$ and T, since these are common for both constituents,

$$\omega = \frac{m_a}{m_v} = \frac{P_v}{R_v}\frac{R_a}{P_a}$$

Now, R_a can be taken as 0.287 kJ/kg K and R_v as $8.314/18 = 0.462$ kJ/kg K, so that

$$\omega = \frac{0.287}{0.462}\frac{P_v}{P_a} = 0.622\frac{P_v}{P_a} \tag{10.11}$$

This defines the amount of water vapour in the air. If there is no water vapour present then the specific humidity is clearly zero. But this is a situation that is never attained in reality. Even in the middle of the Sahara desert there will **always** be a **small amount** of water vapour present and the value of ω will be above zero. The value of ω will vary, depending on the weather conditions prevailing at the time. These are defined in terms of the ambient temperature pressure and the relative humidity. The atmospheric pressure P is the sum of the pressure contributed by the dry air and the pressure contributed by the water vapour:

$$P = P_a + P_v \tag{10.12}$$

For a given atmospheric pressure, as the specific humidity increases, P_v will increase and P_a will decrease.

At a given ambient temperature and pressure there is a maximum value to which ω can be increased. At this maximum value the air is said to be 'saturated' and any attempt to add more water vapour to the air will cause the excess to simply condense out in the form of water droplets. Looking at Figure 10.2, this saturated condition occurs when the ambient temperature equals the saturation temperature of the water vapour. Expressed another way, in terms of the pressure of the water vapour, saturation occurs when

$$P_v = P_{sat}$$

Taking the numerical example given earlier, a water vapour pressure of 1.7 kPa is equivalent to a saturation temperature of 15°C. At an ambient temperature of, say 20°C, the actual temperature is above the saturation temperature and the water vapour is superheated. Since the saturation pressure at 20°C is 2.34 kPa, the actual vapour pressure is below this value and the air **cannot** be saturated. However, if the ambient temperature was reduced to 15°C the air **would** then become saturated. How close the air is to being saturated is defined by means of the relative humidity.

Relative humidity is defined as 'the mass of water vapour present as compared to the mass that would be present if the air was saturated' and is given the symbol ϕ:

$$\phi = \frac{m_v}{m_{sat}} \tag{10.13}$$

From the equation of state for a perfect gas, it can be shown that

$$\frac{m_v}{m_{sat}} = \frac{P_v}{P_{sat}}$$

and

$$\phi = \frac{P_v}{P_{sat}} \tag{10.14}$$

Values of relative humidity are generally quoted on a percentage basis.
Appendix B1 quotes values of saturation pressure for a limited range of air temperatures.

Example 10.4

Air at an ambient temperature of 20°C and an atmospheric pressure of 101 kPa, has a relative humidity of 70%. Calculate the specific humidity under these conditions.

Analysis — from Appendix B1 at 20°C,

$$P_{sat} = 2.337 \text{ kPa}$$

From equation (10.14)

$$\phi = \frac{P_v}{P_{sat}}$$

$$0.7 = \frac{P_v}{2.337}$$

$$P_v = 1.636 \text{ kPa}$$

Substituting in equation (10.11)

$$\omega = 0.622 \frac{P_v}{P_a}$$

Now,

$$P_a = P - P_v$$
$$= 101 - 1.636$$
$$= 99.364 \text{ kPa}$$

Therefore,

$$\omega = 0.622 \frac{1.636}{99.364} = 0.0102 \text{ kg/(kg air)}$$

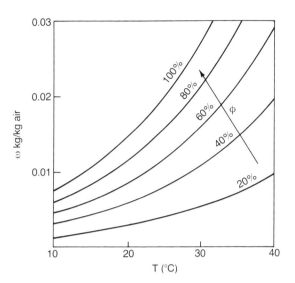

Figure 10.3 Psychrometric chart.

10.5.2 Psychrometric chart

If the calculations, outlined in example 10.4, are carried out for a range of temperatures and relative humidities, it is possible to construct a series of graphs of specific humidity against temperature. These can be combined onto one chart, called the psychrometric chart, as shown in Figure 10.3.

One use of this psychrometric chart is to allow values of specific humidity to be found for any given value of temperature and relative humidity. However, relative humidity is not a property of the mixture of air and water vapour that can be measured directly. It has to be assessed from other measurable quantities, such as the 'wet-bulb' and 'dry-bulb' temperatures.

An ordinary thermometer measures the dry-bulb temperature of the mixture of air and water vapour, so called because the bulb of the thermometer is dry. A wet-bulb thermometer is shown in Figure 10.4. A porous wick surrounds the bulb of the thermometer that is saturated with water. As a flow of unsaturated air crosses the wick there is some evaporation of the water, causing a **lower** temperature to be registered on the wet-bulb thermometer relative to the dry-bulb thermometer. For example, at an ordinary ambient temperature of 20°C, dry-bulb, a wet-bulb temperature of 16°C is equivalent to a relative humidity of 66%.

At a relative humidity of 100%, when the air is saturated, there can be no evaporation from the wet wick and the wet-bulb temperature is then equal to the dry-bulb temperature. This is shown in Figure 10.5, in which the wet-bulb temperatures have been added to the psychrometric chart shown in Figure 10.3.

Adding the wet-bulb temperatures to the psychrometric chart clearly adds to its usefulness. In practice, a psychrometric chart also gives specific enthalpy and specific volume values. Appendix B2 shows such a psychrometric chart as

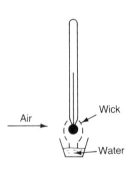

Figure 10.4 Wet-bulb thermometer.

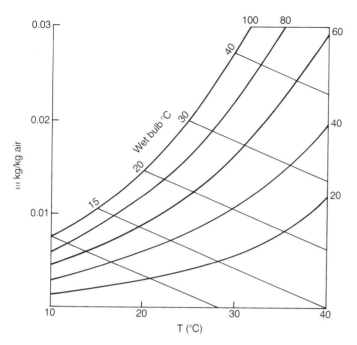

Figure 10.5 Wet-bulb temperatures shown on the psychrometric chart.

published by the Chartered Institution of Building Services Engineers, CIBSE. For the present discussion the chart given in Figure 10.5 illustrates the basic features without involving unnecessary complexity. In this basic form, the psychrometric chart provides a diagram for illustrating the processes involving mixtures of air and water vapour.

10.5.3 Dew-point

For any mixture of air and water vapour there is a saturation temperature, as defined on the T–s diagram shown in Figure 10.2. If the mixture starts with the water vapour in a superheated state, any cooling of the mixture will cause the ambient temperature to eventually reach the saturation temperature. In this state the mixture is **fully** saturated and the relative humidity is 100%. Any slight increase in the amount of water vapour results in the excess moisture condensing out of the mixture. Alternatively, any slight reduction in the temperature below the saturation temperature will cause some of the water vapour to condense. Therefore, any condensation process **starts** at the saturation temperature.

There is plenty of visual evidence of condensation taking place. In cold weather, condensation takes place on the inside of the windows of a warm room. This is because the glass has been cooled down to below the saturation temperature of the air in the room. Another example is the formation of 'dew' on grass in the early morning, before the increase in air temperature causes it to evaporate. The dew is the water that has condensed from the atmosphere overnight when the ground temperature dropped below the saturation temperature.

The 'dew-point' is defined as the temperature at which dew will start to appear. In other words, it is the temperature at which condensation will start for a given mixture of air and water vapour. Since condensation starts at the saturation temperature, it follows that the dew-point is equal to the saturation temperature of the mixture.

Assuming that the pressure of an air and water vapour mixture remains constant, a cooling process that brings about condensation is shown on the psychrometric chart in Figure 10.6. The mixture starts at state 1 with a given temperature T and specific humidity ω. As the mixture is cooled the temperature drops but the specific humidity remains constant. If the temperature drops sufficiently low, the mixture reaches saturation temperature at state 2. At this state the temperature of the mixture is at the dew-point and a further slight reduction of the temperature will cause condensation to take place.

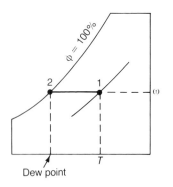

Figure 10.6 Cooling process.

Example 10.5
Air at sea level has a temperature of 25°C and a relative humidity of 74%. Assuming that the air temperature drops by 6.5 K for each 1000 m increase in height, estimate the altitude at which clouds will start to appear. Assume the air to be at constant pressure.

Process diagram — Figure 10.6.

Analysis — clouds form when moisture condenses out of the atmosphere, i.e. when the air temperature reaches the dew-point.
At sea level, from Appendix B1, at 25°C

$$P_{sat} = 3.166 \text{ kPa}$$

From equation (10.14)

$$\phi = \frac{P_v}{P_{sat}}$$

$$0.74 = \frac{P_v}{3.166}$$

$$P_v = 2.34 \text{ kPa}$$

This vapour pressure is approximately equal to the saturation pressure at 20°C. Therefore,

$$\text{dew-point} \approx 20°C$$

$$\text{altitude} \approx \frac{25 - 20}{6.5} \times 1000 = 769 \text{ m}$$

10.6 PSYCHROMETRIC PROCESSES

The psychrometric chart, as shown in Figure 10.5, provides numerical data regarding mixtures of air and water vapour. In addition, it also provides a diagram that can be used to visualize various psychrometric processes. One such process has already been portrayed in Figure 10.6, the cooling process. Since a cooling process does not cause any change to the relative mass of the water vapour with relation to the air, the specific humidity remains constant. However, as the air temperature is reduced the vapour pressure approaches the saturation pressure and the relative humidity increases.

The basic psychrometric processes are shown in Figure 10.7 and consist of heating, cooling, humidifying and dehumidifying. Both heating and cooling take place at **constant** specific humidity, whereas both humidifying and dehumidifying involve **changes** to the specific humidity.

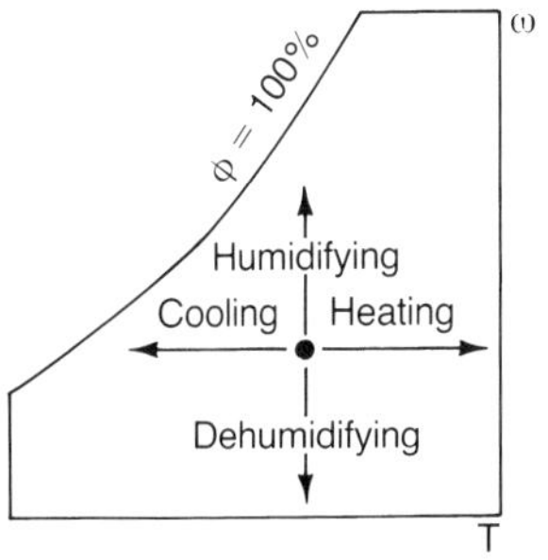

Figure 10.7 Psychrometric processes.

10.6.1 Heating

One of the most popular forms of heating within the home is by means of central heating. Such systems use a central boiler that circulates hot water to individual radiators in each room of the house. Prior to the widespread application of central heating, heating was achieved through individual fires or portable heaters. One effect of central heating is to raise the temperature of the whole air contained within the house.

Two criticisms of central heating are that the air tends to feel 'dry' and that it causes wooden furniture to crack. Both of these are associated with a reduction in relative humidity. As the air temperature increases, the specific humidity remains constant but the air moves to the right on the psychrometric chart, as shown in Figure 10.7, to a region of reduced relative humidity.

Although specific humidity is an important criterion when analysing mixtures of air and water vapour, it is relative humidity that determines the comfort of a human body. This is because the human body can be viewed as a heat engine that has to reject waste heat to the surroundings. The rejection of the waste heat is by perspiration, the evaporation of sweat in order to cool the body. A human body feels comfortable when it can freely reject the waste heat. However, it feels uncomfortable if it has either difficulty in rejecting the waste heat, when the relative humidity is high, or rejects excess heat when the relative humidity is low. In the latter case, it results in the body feeling dry and needing liquid replenishment to compensate for the moisture lost by perspiration.

There are no hard and fast rules as to the required temperature and relative humidity for a human to feel comfortable. It depends on age and the individual person, but an ambient temperature range of 20–25°C and a relative humidity of 40–60% would generally be considered a comfortable environment.

Example 10.6
If the outside conditions are 10°C and a relative humidity of 70%, calculate the relative humidity in the living room of a house that is heated to 25°C.

Process diagram

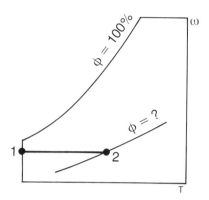

Analysis — from Appendix B1, at 10°C

$$P_{sat} = 1.227 \text{ kPa}$$

From equation (10.14)

$$\phi = \frac{P_v}{P_{sat}}$$

$$0.7 = \frac{P_v}{1.227},$$

$$P_v = 0.859 \text{ kPa}$$

At 25°C

$$P_{sat} = 3.166 \text{ kPa}$$

Therefore,

$$\varphi \text{(living room)} = \frac{0.859}{3.166} = 0.271 \text{ i.e. } 27\%$$

Note — under these conditions some form of humidification would be required in the living room to ensure a comfortable environment.

10.6.2   Dehumidification

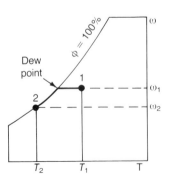

Figure 10.8 Dehumidification process.

Dehumidification is the process of reducing the amount of water vapour in the air. In order to achieve this, it is necessary to reduce the value of specific humidity. Unfortunately, it is not possible to achieve this through a simple process, as illustrated in Figure 10.7. The reduction of specific humidity requires some of the water vapour to be removed and this can be achieved by condensation. A typical dehumidification process is shown in Figure 10.8.

The moist air starts at state 1, and is required to be dehumidified to specific humidity ω_2. To achieve condensation of the water vapour from the air at state 1, it is necessary to cool the mixture down to its dew-point. The condensation process starts when the mixture reaches saturation condition at $\phi = 100\%$. Continuing condensation of the water vapour requires the temperature to be reduced below the dew-point until it reaches the wet-bulb temperature associated with ω_2.

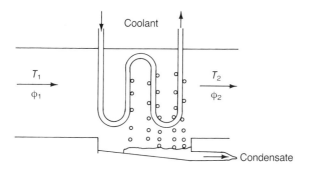

Figure 10.9 Dehumidifier.

Since the dehumidification process is basically a cooling process, a dehumidifier consists of a heat exchanger in which the surface is maintained at a low temperature, below T_2. A schematic diagram of a dehumidifier is shown in Figure 10.9.

Air enters the heat exchanger and is cooled down to its dew-point. Thereafter water condenses on the surface of the tubes and flows downwards, under the action of gravity, to be caught in a condensate trap. The condensate then either runs to a drain or is collected in a storage container. Portable dehumidifiers collect the condensate in containers that have to be emptied on a regular basis. To achieve the necessary cooling, the fluid inside the tubes of the heat exchanger must be at a temperature well below the outlet temperature T_2. For large-scale permanent dehumidifiers the coolant can be chilled water. For portable dehumidifiers the coolant is refrigerant and the heat exchanger forms the evaporator of a small refrigeration system.

Example 10.7
Air enters a dehumidifier with a temperature of 20°C and a relative humidity of 80%. If a quarter of the water content is to be removed, estimate the temperature of the air leaving the dehumidifier. Assume the process to be at a constant pressure of 100 kPa.

Conceptual model — Figure 10.9.

Process diagram — Figure 10.8.

Analysis — at state 1, $T_1 = 20$°C and $\phi_1 = 0.8$.
From Appendix B1

$$P_{sat} = 2.337 \text{ kPa}$$

From equation 10.14

$$\phi = \frac{P_v}{P_{sat}}$$

$$0.8 = \frac{P_{v_1}}{2.337}$$

$$P_{v_1} = 1.87 \text{ kPa}$$

Substituting in equation (10.11)

$$\omega = 0.622 \frac{P_v}{P_a}$$

Now

$$P_{a_1} = P - P_{v_1}$$
$$= 100 - 1.87 = 98.13 \text{ kPa}$$

Therefore,

$$\omega_1 = 0.622 \frac{1.87}{98.13} = 0.0119 \text{ kg/(kg air)}$$

At state 2, a quarter of the water has been removed:

$$\omega_2 = 0.75\, \omega_1 = 0.75 \times 0.0119$$
$$= 0.0089 \text{ kg/(kg air)}$$

From equation (10.11)

$$0.0089 = 0.622 \frac{P_{v_2}}{100 - P_{v_2}}$$

Therefore

$$P_{v_2} = 1.415 \text{ kPa}$$

Since the condition at state 2 is saturated,

$$P_{sat_2} = 1.415 \text{ kPa}$$

From Appendix B1 it will be seen that this value of saturation is equivalent to a temperature T_2 between 10 and 15°C. Interpolating between these two values

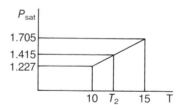

$$T_2 = 10 + \frac{1.415 - 1.227}{1.705 - 1.227}(15 - 10) = 11.97°C \text{ i.e. } 12°C$$

Note — the interpolation is based upon the assumption that the pressure varies **linearly** with temperature. This is not quite true, a saturation pressure of 1.415 kPa is equivalent to a temperature of 12.1°C, but the answer of 12°C is sufficiently accurate for the purpose of this example.

10.6.3 Air Conditioning

Air conditioning is necessary in tropical climates where both the ambient temperature and relative humidity are too high for comfort. To achieve a comfortable living, and working, environment it is necessary to reduce the moisture content of the air. This can be done by a dehumidification process, as outlined in the previous section. However, such a process results in air at a low temperature and a relative humidity of 100%. To ensure that the air is then changed to a more comfortable condition, it must be heated. Therefore, air conditioning consists of two processes, dehumidification followed by heating, as shown in Figure 10.10.

Air starts at state 1 with a high ambient temperature T_1 and high relative humidity ϕ_1 Specific humidity is reduced by dehumidifying to state 2. The air is then heated, at constant specific humidity, until a satisfactory temperature T_3 and relative humidity ϕ_3 are achieved. The result of the air conditioning processes are to reduce the temperature and relative humidity. As portrayed in Figure 10.10,

$$T_3 < T_1$$
$$\phi_3 < \phi_1$$

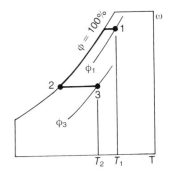

Figure 10.10 Air conditioning processes.

10.6.4 Humidification

With the generally humid conditions that prevail in Britain, it would seem unnecessary to consider humidification. Nevertheless, humidification is a very important topic since some devices, typically 'tumble dryers' and 'cooling towers', can only function by increasing the moisture content of the air flowing through the system.

A tumble dryer is a device for drying wet clothes and linen, by blowing warm air through the items as they rotate in a drum. The rotation ensures that the items of clothing and linen are in intimate contact with the warm air as it passes through the drum. The major energy input to such a dryer is required to heat the air before entering the drum. This energy input is achieved by either using an electric resistance heater or a gas fired heater. As the warm air passes through the items of clothing and linen it absorbs moisture and thereby ensures that drying takes place. As a result of the drying process, the air leaves the tumble dryer at a higher temperature and relative humidity than when it entered.

Figure 10.11 shows the process that takes place within a tumble dryer. Air enters at temperature T_1 and relative humidity ϕ_1. It is heated to an increased temperature but lower relative humidity. The warm air enters the dryer and absorbs moisture. This increases both the specific humidity and relative humidity. However, the evaporation of the water results in humidification of the air but a reduction in the temperature, with a final condition of T_2 and ϕ_2. It is not necessary to worry about the intermediate heating and humidification processes, the whole drying process can be defined in terms of the inlet and outlet conditions at states 1 and 2 respectively.

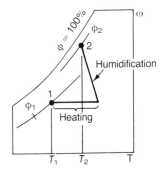

Figure 10.11 Drying process.

The rate at which the moisture is absorbed is the difference between the mass flow rate of water vapour leaving $\dot{m}_{v_2}$, and the mass flow of water vapour entering $\dot{m}_{v_1}$. However,

$$\dot{m}_{v_2} = \omega_2 \, \dot{m}_a$$

and

$$\dot{m}_{v_1} = \omega_1 \, \dot{m}_a$$

where $\dot{m}_a$ is the mass flow rate of the dry air. Therefore,

$$\dot{m}_{v_2} - \dot{m}_{v_1} = \dot{m}_v \, (\omega_2 - \omega_1) \tag{10.15}$$

A cooling tower is another device that uses the humidification of air to achieve the required performance. However, in the case of a cooling tower the aim is not to cause evaporation to reduce moisture levels, but to use the heat transfer necessary for evaporation for cooling. Cooling towers are very familiar devices; they dominate the landscape where inland power stations are situated, as shown in Figure 10.12. Their purpose is to cool the water used for condensing the steam in the thermal power plant. The high quantity of waste heat has to be dissipated to the atmosphere and the most economic way of achieving this is by means of cooling towers.

Water enters the cooling tower at a temperature **well** above ambient and falls under the action of gravity down to a collection unit. Air flows in a counter

Figure 10.12 Cooling towers in operation.

flow direction to the water, causing some evaporation which cools the bulk of the water. The air gains both temperature, and moisture, and is lighter than the air surrounding the tower, so it flows upwards by natural draught. The condition of the air leaving the tower is at, or near, saturation. This is evident when a cooling tower is seen in operation. The moisture in the air leaving at the top of the tower shows as a visible cloud of vapour, very much like smoke leaving a chimney.

There will be a loss of water due to evaporation, as defined by equation (10.15). This loss has to be made up by the addition of water to ensure that the mass flow rate of the cooling water through the condenser is maintained at the correct level.

Although the cooling towers associated with power stations are the most obvious examples of these devices, cooling towers are also employed in general applications. Cooling towers are widely employed for processes in industry, although such towers tend to be much smaller than those used for power stations. The smaller towers are constructed of wood or sheet metal and sometimes have a fan to help the flow of air. The large natural-draught towers employed for power stations are constructed of reinforced concrete.

Example 10.8
Air enters a tumble dryer at 20°C and a relative humidity of 65%. If the air leaves the dryer in a saturated condition at 45°C, calculate the mass flow rate of dry air through the dryer in order to evaporate 1 kg of water in 40 minutes. Assume the process to take place at a constant atmospheric pressure of 100 kPa.

Conceptual model

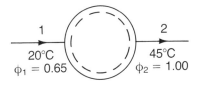

Process diagram — Figure 10.11.

Analysis — at state 1 $T_1 = 20°C$ and $\phi_1 = 0.65$.
From Appendix B1, $P_{sat_1} = 2.337$ kPa.
From equation (10.14)

$$\phi = \frac{P_v}{P_{sat}}$$

$$0.65 = \frac{P_{v1}}{2.337}$$

$$P_{v1} = 1.519 \text{ kPa}$$

Substituting in equation (10.11)

$$\omega = 0.622 \frac{P_v}{P_a}$$

Now

$$P_{a_1} = P - P_{v_1} = 100 - 1.519 = 98.48 \text{ kPa}$$

Therefore,

$$\omega_1 = 0.622 \frac{1.519}{98.48} = 0.0096 \text{ kg/(kg air)}$$

At state 2 $T_2 = 45°C$ and $\phi_2 = 1.0$.
From Appendix B1, $P_{sat_2} = 9.577$ kPa.
Now

$$P_{a_2} = P - P_{v_2} = 100 - 9.577 = 90.42 \text{ kPa}$$

and

$$\omega_2 = 0.622 \frac{9.577}{90.42} = 0.0659 \text{ kg/(kg air)}$$

Therefore, the amount of water evaporated is

$$\dot{m}_{v_2} - \dot{m}_{v_1} = \frac{1}{40 \times 60} = 0.000417 \text{ kg/s}$$

From equation (10.12)

$$\dot{m}_{v_2} - \dot{m}_{v_1} = \dot{m}_a(\omega_2 - \omega_1)$$
$$0.000417 = \dot{m}_a(0.0659 - 0.0096)$$

Therefore,

$$\dot{m}_a = 0.0074 \text{ kg/s}$$

Example 10.9
Air enters a cooling tower with a volumetric flow rate of 10 m³/s. The air enters with a temperature of 15°C and relative humidity of 50%, and leaves at 30°C in a saturated condition. Calculate the rate of evaporation assuming that the process takes place at a constant pressure of 101 kPa. Take R for air as 0.287 kJ/kg K.

Conceptual model

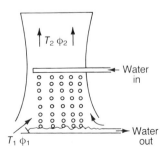

Process diagram

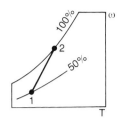

Analysis — at state 1 $T_1 = 15°C$ and $\phi_1 = 0.5$.
From the equation of state, (4.6),

$$v_1 = \frac{RT_1}{P_1} = \frac{0.287 \times 288}{101} = 0.818 \text{ m}^3/\text{kg}$$

Mass flow rate of air:

$$m_a = 10/0.818 = 12.22 \text{ kg/s}$$

From Appendix B1 $P_{sat_1} = 1.705$ kPa
From equation (10.14):

$$\phi = \frac{P_v}{P_{sat}}$$

$$0.5 = \frac{P_{v_1}}{1.705}$$

$$P_{v_1} = 0.853 \text{ kPa}$$

Substituting in equation (10.11)

$$\omega = 0.622 \frac{P_v}{P_a}$$

Now

$$P_{a_1} = P - P_{v_1} = 101 - 0.853 = 100.15 \text{ kPa}$$

Therefore,

$$\omega_1 = 0.622 \frac{0.853}{100.15} = 0.0053 \text{ kg/(kg air)}$$

At state 2 $T_2 = 30°C$, $\phi_2 = 1.0$
From Appendix B1 $P_{sat_2} = 4.241$ kPa
Now

$$P_a = P - P_{v_2} = 101 - 4.241 = 96.76 \text{ kPa}$$

and

$$\omega_2 = 0.622 \times \frac{4.241}{96.76} = 0.0273 \text{ kg/(kg air)}$$

The rate of evaporation is found from equation (10.15):

$$\dot{m}_{v_2} - \dot{m}_{v_1} = \dot{m}_a(\omega_2 - \omega_1)$$
$$= 12.22 \times (0.0273 - 0.0053)$$
$$= 0.27 \text{ kg/s}$$

10.7 PSYCHROMETRIC ENERGY EQUATION

All the psychrometric processes described in the preceding sections can be considered to be steady flow processes that take place within some form of heat exchanger. Even a tumble dryer can be considered as a heat exchanger in which the air has to be heated before entering the drum. It is possible to analyse the rate of heat transfer that takes place within a psychrometric process by applying the steady flow energy equation, defined in Chapter 5.

A typical psychrometric process can be modelled as shown in Figure 10.13. The device can be considered as an open system into which a mass flow rate of air $\dot{m}_a$ enters at state 1. The same mass flow rate of air leaves the system at state 2. In addition, a mass flow rate of water vapour $\dot{m}_{v_1}$ enters the system and a mass flow rate of water vapour $\dot{m}_{v_2}$ leaves the system. For processes where humidification or dehumidification takes place, $\dot{m}_{v_2}$ will be different to $\dot{m}_{v_1}$.

Assuming that there is no work done during the process and that the change of velocity across the system is negligible, the process shown in Figure 10.13 can be analysed using the steady flow energy equation as a rate equation. From equation (5.10) it can be shown that the rate of change of energy across the system is given by

$$Q + \dot{m}_a h_{a_1} + \dot{m}_{v_1} h_{v_1} = \dot{m}_a h_{a_2} + \dot{m}_{v_2} h_{v_2} \tag{10.16}$$

This can be expressed as

$$Q + \dot{m}_a(h_{a_1} + \omega h_{v_1}) = \dot{m}_a(h_{a_2} + \omega_2 h_{v_2}) \tag{10.17}$$

A study of the psychrometric chart given in Appendix B2, indicates that enthalpy values are quoted as specific enthalpies, kJ/kg. The 'kg' in this context means one kg of air. So the enthalpy values quoted apply to **1 kg of dry air** and should really be expressed as kJ/(kg air).

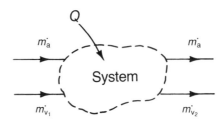

Figure 10.13 Steady flow psychrometric process.

This provides a very useful method of analysing the energy changes during a psychrometric process, as equation (10.17) can be expressed:

$$Q + \dot{m}_a h_1 = \dot{m}_a h_2$$

or

$$Q = \dot{m}_a (h_2 - h_1) \qquad (10.18)$$

where the enthalpy values h_1 and h_2 represent the total enthalpy of a mixture of dry air and water vapour, quoted with relation to the mass flow rate of dry air.

Values of h_1 and h_2 may be found directly from the psychrometric chart shown in Appendix B2. Alternatively, values of enthalpy can be calculated directly in the following manner.

The enthalpy of a mixture of dry air and water vapour can be found using the relationship

$$h = h_a + \omega h_v$$

Within the range of temperatures considered in psychrometric processes, dry air can be assumed to be a perfect gas with a value of C_P very close to 1 kJ/kg K. Taking the datum for the enthalpy values to be 0°C, the enthalpy of the dry air can be expressed as

$$h_a = C_P (T - 0) \approx T \qquad (10.19)$$

where T is the temperature of the mixture in °C.

The enthalpy of the water vapour can be taken as being equal to the enthalpy of the saturated vapour at the same temperature, T. The enthalpy of water vapour at 0°C is approximately 2501 kJ/kg. Assuming the water vapour to behave as a perfect gas, the value of C_P can be taken as 1.82 kJ/kg K and the enthalpy of the water vapour expressed in the form

$$h_v = 2501 + 1.82\, T \qquad (10.20)$$

Although this relationship is only approximate, the values are very close to those quoted in tables. For the temperature range 10–50°C the maximum difference is 0.6 kJ/kg, i.e. an error of less than 0.03%.

Combining equations (10.19) and (10.20) gives the following relationship for the enthalpy of a mixture of dry air and water vapour:

$$h = T + \omega (2501 + 1.82\, T) \qquad (10.21)$$

where h is expressed in kJ/(kg air).

Example 10.10

Calculate the rate of heat transfer to the tumble dryer defined in example 10.8.

Conceptual model

Process diagram — considering the process to be defined by the condition of the air entering and leaving the dryer, the process can be visualized as

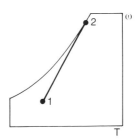

Analysis — At state 1, from problem 10.8, $T_1 = 20\ °C$ and $\omega_1 = 0.0096\ kg/(kg$ air).
From equation (10.21)

$$h_1 = T_1 + \omega_1\ (2501 + 1.82\ T_1)$$
$$= 20 + 0.0096\ (2501 + 1.82 \times 20)$$
$$= 44.36\ kJ/(kg\ air)$$

At state 2, from problem 10.8, $T_2 = 45\ °C$ and $\omega_2 = 0.0659 kg/(kg\ air)$.
From equation (10.21)

$$h_1 = 45 + 0.0659\ (2501 + 1.82 \times 45)$$
$$= 215.21\ kJ/(kg\ air)$$

The rate of heat transfer can be found using equation (10.18):

$$Q = \dot{m}_a(h_2 - h_1)$$

From problem 10.8, $\dot{m}_a = 0.0074\ kg/s$ and

$$Q = 0.0074\ (215.21 - 44.36)$$
$$= 1.26\ kW$$

Note — it is not possible to solve this problem using the psychrometric chart given in Appendix B2 as the conditions at state 2 are outside the range of the chart.

10.8 SUMMARY

In this chapter the properties and behaviour of mixtures of perfect gases have been discussed.
  The key terms that have been introduced are:

mixtures of gases
Dalton's law of partial pressures
Gibbs–Dalton law
psychrometry

specific humidity
relative humidity
psychrometric chart
dry-bulb temperature
wet-bulb temperature
dew-point
humidifying
dehumidifying
air conditioning
tumble dryer
cooling tower

Key equations that have been introduced.

For mixtures of gases:

$$mR = m_1 R_1 + m_2 R_2 + \ldots \qquad (10.5)$$

$$mC_P = m_1 C_{P_1} + m_2 C_{P_2} + \ldots \qquad (10.8)$$

$$mC_v = m_1 C_{v_1} + m_2 C_{v_2} + \ldots \qquad (10.9)$$

In psychrometry,
   for specific humidity:

$$\omega = 0.622 \, \frac{P_v}{P_a} \qquad (10.11)$$

   for relative humidity:

$$\phi = \frac{P_v}{P_{sat}} \qquad (10.14)$$

   for rate of moisture absorption:

$$\dot{m}_{v_2} - \dot{m}_{v_1} = \dot{m}_a (\omega_2 - \omega_1) \qquad (10.15)$$

   for rate of heat transfer:

$$\dot{Q} = \dot{m}_a (h_2 - h_1) \qquad (10.18)$$

   where

$$h = T + \omega \, (2501 + 1.82 \, T) \qquad (10.21)$$

PROBLEMS

Mixtures of Gases

1. A mixture of gases consists of 40% nitrogen (N_2), 25% hydrogen (H_2) and 35% carbon dioxide (CO_2) by mass. If the mixture is at a pressure of 100 kPa and a temperature of 17°C, calculate:

(a) the partial pressure of each constituent gas;
(b) the specific volume of the mixture.

2. A container having a volume of 1 m^3 has oxygen (O_2) at 50°C and 400 kPa. In an isothermal process, nitrogen (N_2) is pumped in until the pressure reaches 600 kPa. What is the final mixture composition based on mass?

3. A cylinder and frictionless piston contains 1 kg of a mixture of gases. The mixture consists of 50% carbon dioxide (CO_2), 30% oxygen (O_2) and 20% hydrogen (H_2) by mass. During a reversible constant pressure process the temperature of the gas increases from 50°C to 120°C. Calculate the heat transferred during the process.
   Assume

C_P (CO_2) = 0.85 kJ/kg K
C_P (O_2) = 0.92 kJ/kg K
C_P (H_2) = 14.2 kJ/kg K

4. A mixture of gases consisting of 60% nitrogen (N_2) and 40% carbon dioxide (CO_2) by mass, expands through a turbine from a temperature of 1200 K and a pressure of 800 KPa, down to an exhaust pressure of 160 kPa. Calculate the specific work output assuming the expansion to be a reversible adiabatic process. Assume

C_P (N_2) = 1.05 kJ/kg K
C_P (CO_2) = 0.94 kJ/kg K

Psychrometry

5. For air at a temperature of 25°C and atmospheric pressure of 101 kPa, calculate:

(a) the specific humidity if the relative humidity is 70%;
(b) the relative humidity if the specific humidity is 0.0119 kg/(kg air).

6. Air at a temperature of 30°C and a relative humidity of 80%, enters a dehumidifier. If a fifth of the moisture is removed during the dehumidification process, estimate the temperature of the air leaving. Assume the process to be at a constant pressure of 101 kPa.

7. Air enters a cooling tower with a temperature of 15°C and relative humidity of 60%. If the air leaves at 30°C in a saturated condition, find the amount of water lost for each kg of air flowing through the tower. Assume the cooling process to be a constant pressure of 101 kPa.

8. Air enters a tumble dryer at a temperature of 20°C and relative humidity of 70%. Find the mass flow rate of air required to remove 1 kg of water in 30 minutes, if the air leaving the dryer is fully saturated at a temperature of 40°C. Assume the drying process to be performed at an atmospheric pressure of 100 kPa.

9. Air enters a cooling tower at the rate of 12 kg/s with a temperature of 20°C and relative humidity of 50%. If the air leaves at 35°C in a saturated condition, find the rate of heat transfer during this process. Assume the atmospheric pressure to be 100 kPa.

11 | Combustion

11.1 AIMS

- To discuss fuels in terms of their physical state:

  solid
  liquid
  gaseous

- To define combustion as a process complying with the principle of conservation of mass.
- To derive the chemical equations for combustion with oxygen.
- To discuss combustion with oxygen within rocket engines.
- To derive the chemical equations for combustion with air.
- To define:

  air–fuel ratio
  percentage excess air.

- To discuss the air–fuel ratio of internal combustion engines.
- To define the enthalpy of formation for a compound.
- To use the enthalpy of formation to calculate the enthalpy of combustion for a steady flow process.
- To apply the steady flow energy equation to adiabatic combustion processes.

11.2 FUELS

The analysis of power cycles, whether operating with steam or air as the working fluid, requires a heat input to the cycle. Apart from some nuclear power stations, the cycles operate with a heat input achieved by the combustion of fuel. Combustion is a chemical reaction by which fuel is combined with a source of oxygen to release thermal energy. The source of oxygen is usually the air in the atmosphere. Before discussing the chemical reaction involved in combustion it is necessary to briefly discuss the composition of fuels.

Fuels can be classified by their physical state under normal conditions and are defined as either solid, liquid or gaseous. Most power stations operating on steam power cycles use coal as a solid fuel, although there is a move towards the use of natural gas. Engines used for transportation and aircraft propulsion operate with liquid fuels. Industrial gas turbine engines operate with either liquid fuels or natural gas.

The source of all fuels is sunlight, the solar energy reaching the surface of the earth. Renewable solid fuels such as wood and peat require solar energy for their growth. Coal, crude oil and natural gas represent fossil fuels, so-called because their creation depended on the plants and creatures growing on the earth many millions of years ago. Since the fossil fuel reserves were created so long ago, they represent a finite quantity that is non-renewable. Unfortunately, the supply of renewable fuels is limited and mankind continues to use fossil fuel with the inevitable depletion of their reserves. One task facing engineers in the 21st century is the development of alternative energy sources.

Solid fuels include wood, peat and coal. The compositions of these fuels vary from one sample to another, but Table 11.1 proves a general guide.

The data in Table 11.1 are based on the assumption that the fuel is free of moisture and ash. In practice, coal contains a small proportion of each. Wood and peat contain a higher proportion of moisture, about 20% by mass.

Liquid fuels are generally distilled from crude oil, as shown in Figure 11.1, although it is possible to derive some liquid fuels from coal. The most volatile elements of the crude oil vaporize first during the distillation process, forming petrol. In order, the less volatile elements form diesel fuel and fuel oils. All of these consist mainly of hydrogen and carbon, forming a range of 'hydrocarbon fuels'. The composition of a particular liquid fuel depends on the oil field from which the crude oil is obtained and on the refining process. A general guide as to the composition of various liquid fuels is given in Table 11.2.

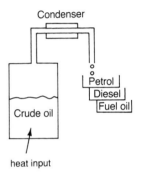

Figure 11.1 Distillation of oil.

Table 11.1 Properties of solid fuels

Property composition (% mass)	Wood	Peat	Bituminous coal	Anthracite
Carbon	50	60	85	94
Hydrogen	6	6	5	3
Oxygen	43	32	6	2
Energy content (kJ/kg)	16 000	18 000	32 000	34 000

Table 11.2 Properties of liquid fuels

Property Composition (% mass)	Petrol	Diesel oil	Fuel oil
Carbon	85	86	86
Hydrogen	14	13	12
Energy content (kJ/kg)	45 000	44 000	43 000

Table 11.3 Properties of gaseous fuels

Property Composition (% volume)	Natural gas	Coal gas
Methane	93	20
Carbon monoxide	1	18
Hydrogen	–	50
Energy content (kJ/m^3)	35 000	19 000

With the widespread availability of natural gas, it is the most popular type of gaseous fuel for use both in the home and in industry. Prior to the discovery of natural gas in the North Sea gas fields, gas was produced from coal. A comparison of these two fuels is presented in Table 11.3, with the data expressed in terms of volume. The major constituent of natural gas is methane (CH$_4$) and for simplicity in modelling combustion with natural gas it is taken to be 100% methane.

From a study of the tables it will be seen that all fuels have carbon and hydrogen as the main combustible elements within their composition. Although not mentioned above, some fuel oils and coals also contain sulphur. This is mainly as an impurity, but it does combine with oxygen to form sulphur dioxide during combustion. As the sulphur content rarely exceeds 1 to 2% of the fuel, it plays a negligible part in the energy release during combustion and is ignored in the analysis of combustion processes. Unfortunately, the sulphur dioxide combines with water vapour in the combustion gases to form a dilute solution of sulphuric acid. This can create problems, of which the environmental damage due to acid rain is one.

11.3 COMBUSTION PROCESSES

To achieve combustion, a fuel must be mixed with a source of oxygen. The source of oxygen is called the 'oxidant'. Combustion takes place within a combustion chamber and, for a steady-flow combustion process, this can be visualized as shown in Figure 11.2.

Within the combustion chamber the fuel and oxidant combine in a chemical reaction that releases thermal energy. Before combustion, the fuel and oxidant

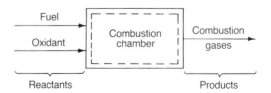

Figure 11.2 A steady-flow combustion process.

are called the 'reactants'. The hot gases resulting from combustion are called the 'products' of combustion.

For combustion to take place the fuel and oxidant must enter the combustion chamber in the correct proportions. This can be assessed from the composition of the fuel and oxidant. The necessary chemical relationships are discussed below. In addition, the fuel must be at a high enough temperature for ignition to take place. Approximate ignition temperatures are listed below:

Petrol   550 K
Coal   700 K
Hydrogen   850 K
Natural gas   900 K

Ignition is started by means of a flame or spark. Once steady flow combustion has started, as in a boiler or gas turbine engine, the ignitor can be turned off because the combustion process is continuous and self-sustaining. However, in cyclic combustion, as in the cylinder of a petrol engine, a spark is required to start combustion during each cycle.

Combustion processes must comply with the principle of conservation of mass. In other words, the mass flow rate of the products leaving the combustion chamber must be equal to the mass flow rate of the fuel and oxidant entering. The conservation of mass is the basis for analysing the chemical reactions described below.

11.3.1 Chemical reactions

The main combustible constituents of fuels are carbon (C) and hydrogen (H_2). When combined with oxygen (O_2) during a combustion process the reactions are defined by the following chemical equations:

$$C + O_2 \rightarrow CO_2 \tag{11.1}$$
$$H_2 + \tfrac{1}{2}O_2 \rightarrow H_2O \tag{11.2}$$

Looking at the combustion of carbon on its own, the above chemical equation is based on the assumption that there is sufficient oxygen for complete combustion. If the amount of oxygen is less than sufficient, some of the carbon will be changed to carbon monoxide (CO) which itself is a fuel and can be combined with additional oxygen to form carbon dioxide (CO_2). As a basis for all the following discussion it is assumed that all combustion processes are complete.

The above chemical equation for the combustion of carbon has carbon and oxygen as the reactants, with carbon dioxide as the product of the combustion process. In addition, it complies with the principle of conservation of mass. The equation states that one molecule of carbon combines with one molecule of oxygen to form one molecule of carbon dioxide. This can be expressed in terms of the molecular unit, the 'kmol', as follows:

$$C \quad + O_2 \quad \rightarrow CO_2$$
$$1 \text{ kmol} + 1 \text{ kmol} = 1 \text{ kmol}$$

The actual mass of a substance contained in a kmol is equal to M, the molecular weight of the substance, as explained in section 1.9*. As a result, the reaction can be defined in terms of mass as follows:

$$C \quad + O_2 \quad \rightarrow CO_2$$
$$1 \text{ kmol} + 1 \text{ kmol} = 1 \text{ kmol}$$
$$12 \text{ kg} + 32 \text{ kg} = 44 \text{ kg}$$

From this, the mass of oxygen required for the complete combustion of 1 kg of carbon can be found, i.e.

$$32/12 = 2.667 \text{ kg O}_2 \text{ /kg C}$$

Using the same type of argument, the reaction of hydrogen with oxygen to form water vapour can be defined as follows:

$$H_2 \quad + \tfrac{1}{2}O_2 \quad \rightarrow H_2O$$
$$1 \text{ kmol} + \tfrac{1}{2} \text{ kmol} = 1 \text{ kmol}$$
$$2 \text{ kg} + 16 \text{ kg} = 18 \text{ kg}$$

These two chemical equations and the associated masses, are the building blocks for all combustion processes.

Example 11.1

Butane is a hydrocarbon fuel defined by the formula C_4H_{10}. Calculate the mass of oxygen required for the complete combustion of 1 kg of fuel.

Conceptual model

Analysis – assuming complete combustion the chemical equation for this combustion process can be written as

$$C_4H_{10} + aO_2 \rightarrow bCO_2 + cH_2O$$

where a, b and c are unknowns, defining the number of kmol of each substance. The evaluation of three unknowns requires three equations and these can be evaluated by balancing the number of atoms of each element on both sides of the equation:

for carbon 4 $= b$
for hydrogen $10 = 2c$
for oxygen $2a = 2b + c$

It follows that

$$b = 4$$
$$c = 5$$
$$a = 6.5$$

---

* As a reminder, the useful approximate molecular weights are:
  $C = 12$; $H_2 = 2$; $O_2 = 32$; $N_2 = 28$.

The full chemical equation becomes

$$C_4H_{10} \quad + 6.5O_2 \qquad \rightarrow 4CO_2 \quad + 5H_2O$$
$$1 \text{ kmol} \quad + 6.5 \text{ kmol} \quad = 4 \text{ kmol} \quad + 5 \text{ kmol}$$
$$1(58) \text{ kg} + 6.5(32) \text{ kg} = 4(44) \text{ kg} + 5(18) \text{ kg}$$

Since the mass equation balances, it provides a check on the correctness of the chemical equation. Therefore,

$$\text{mass of } O_2/\text{kg } C_4 H_{10} = \frac{6.5 \times 32}{58} = 3.59 \text{ kg}$$

Note – the above derivation of the chemical equation has been given in full as the same principle can be applied to any fuel, not just a hydrocarbon.

Example 11.2

Coal has the composition carbon 84%, hydrogen 5%, water 4% and ash 7%, by mass.
   Calculate the mass of oxygen required for the combustion of each unit mass of coal.

Conceptual model

Analysis – assume that the water and ash content of the coal play no part in the chemistry of the combustion process. Each 1 kg of coal contains 0.84 kg of C. From equation (11.1)

$$C \quad + O_2 \quad \rightarrow CO_2$$
$$12 \text{ kg} + 32 \text{ kg} \quad = 44 \text{ kg}$$

Therefore, 0.84 kg requires

$$0.84 \times \frac{32}{12} = 2.24 \text{ kg } O_2$$

Each 1 kg of coal contains 0.05 kg of H_2. From equation (11.2)

$$H_2 \quad + \tfrac{1}{2}O_2 \quad \rightarrow H_2O$$
$$2 \text{ kg} + 16 \text{ kg} \quad = 18 \text{ kg}$$

Therefore, 0.05 kg requires

$$0.05 \times \frac{16}{2} = 0.4 \text{ kg } O$$

and

$$\text{total mass of } O_2/\text{kg of coal} = 2.24 + 0.4 = 2.64 \text{ kg}$$

11.3.2 Combustion with oxygen

The foregoing discussion considered the combustion of fuels with oxygen alone. In practice, the majority of combustion processes involve oxygen as a constituent of air. However, before considering combustion processes with air, it should be mentioned that there is one thermodynamic system in which combustion can take place with oxygen alone. That is in the combustion chamber of a rocket engine.

A typical liquid fuel rocket system is shown in Figure 11.3. The rocket carries both fuel and oxidant. As a result, it is independent of the atmosphere and can operate in space. The fuel and oxidant are burnt in the combustion chamber before the products flow through the nozzle. To achieve maximum thrust from each unit mass of the products, the exhaust velocity must be as high as possible. A converging–diverging form of nozzle is used to achieve a supersonic exhaust velocity, as explained by Look and Sauer (1988).

Various liquid fuels and oxidants have been used for rocket engines. A brief list is given below.

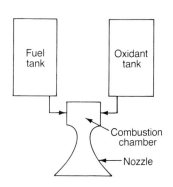

Figure 11.3 Liquid fuel rocket engine.

Fuel	Oxidant
ethyl alcohol, C_2H_6O	oxygen, O_2
kerosine, $C_{12}H_{24}$*	hydrogen peroxide, H_2O_2
hydrogen, H_2	nitric acid, HNO_3

The first large-scale liquid fuel rocket was the V2, used during World War II. It operated with ethyl alcohol as the fuel and liquid oxygen as the oxidant. For small-scale rocket engines, liquid oxygen is inconvenient to use due to its very low boiling point of 90 K. Alternative oxidants that remain liquid at room temperature are hydrogen peroxide and nitric acid, although they give a lower exhaust velocity than liquid oxygen. Space vehicles use a combination of aviation kerosine with liquid oxygen for the lower stages of the rocket and liquid hydrogen with liquid oxygen for the upper stages.

Example 11.3

A rocket engine operates with ethyl alcohol (C_2H_6O) as the fuel and nitric acid (HNO_3) as the oxidant. Calculate the mass of nitric acid required for the complete combustion of 1 kg of fuel.

Conceptual model

Analysis – it can be assumed that the nitrogen atom in a molecule of nitric acid is inert and plays no part in the combustion process. However, it will form a

---

* Kerosine has a complex composition for which $C_{12}H_{24}$ is only an approximation.

separate constituent of the exhaust gases as N_2. The chemical equation for this combustion process can be written as

$$C_2H_6O + aHNO_3 \rightarrow bCO_2 + cH_2O + dN_2$$

where a, b, c and d are unknown numbers of kmol for each substance.

As outlined in example 11.1, these unknowns can be evaluated by balancing the number of atoms of each element on both sides of the equation:

carbon	$2 = b$
hydrogen	$6+a = 2c$
oxygen	$1+3a = 2b+c$
nitrogen	$a = 2d$

It follows that

$$b = 2,$$
$$c = 3 + \tfrac{1}{2}a$$

Substituting in the third equation

$$1 + 3a = 4 + 3 + \frac{1}{2}a$$

$$a = \frac{6}{2.5} = 2.4$$

therefore,

$$c = 3 + \frac{2.4}{2} = 4.2$$

$$d = 1.2$$

The chemical equation becomes:

$$C_2H_6O + 2.4HNO_3 \rightarrow 2CO_2 + 4.2H_2O + 1.2N_2$$
$$1\,kmol + 2.4\,kmol = 2\,kmol + 4.2\,kmol + 1.2\,kmol$$
$$46\,kg + 2.4(63)\,kg = 2(44)\,kg + 42(18)\,kg + 1.2(28)\,kg$$

Therefore,

$$\text{mass of nitric acid/kg fuel} = \frac{2.4 \times 63}{46} = 3.29\,kg$$

11.3.3 Combustion with air

The vast majority of combustion processes rely on air as the means of providing the oxygen. For an analysis of such processes air is assumed to be dry with a simplified composition of 79% nitrogen and 21% oxygen, on a volume basis. Since a volume of gas contains a given number of molecules, irrespective of the gas, the volume proportion also denotes the kmol proportion. Therefore, a kmol of oxygen entering the combustion chamber will be accompanied by $79/21 = 3.76$ kmol of nitrogen.

The chemical equation for the complete combustion of carbon with air can be written as

$$C + O_2 + 3.76N_2 \rightarrow CO_2 + 3.76N_2 \tag{11.3}$$

where the nitrogen is considered to be inert and passes through the combustion process to the combustion products.

In practice, at the very high temperatures encountered in reciprocating petrol engines there is some reaction between the nitrogen and oxygen to form a small proportion of nitric oxide in the exhaust gases. This represents one of the unwanted emissions that can be removed by installing a catalytic converter in the exhaust system. However, such complexities are ignored in the present discussion where nitrogen is considered to be unchanged during the combustion process.

The chemical equation for the complete combustion of hydrogen with air can be written as

$$H_2 + \frac{1}{2}O_2 + \frac{3.76}{2}N_2 \rightarrow H_2O + \frac{3.76}{2}N_2 \tag{11.4}$$

Equations (11.3) and (11.4) form the basis for analysing combustion processes involving air.

___

Example 11.4

A petrol can be approximated as octane, C_8H_{18}. Calculate the mass of air required for the complete combustion of 1 kg of petrol.

Conceptual model

Analysis – the chemical equation for this combustion process can be written as

$$C_8H_{18} + aO_2 + a3.76N_2 \rightarrow bCO_2 + cH_2O + a3.76N_2$$

The values of a, b and c can be evaluated by balancing the number of atoms of each element on both sides of the equation:

carbon   8 $= b$
hydrogen 18 $= 2c$
oxygen   2a $= 2b + c$

It follows that

$b = 8$
$c = 9$
$a = 12.5$

The full chemical equation becomes

$$C_8H_{18} + 12.5O_2 + 47N_2 \rightarrow 8CO_2 + 9H_2O + 47N_2$$

1 kmol + 12.5 kmol + 42 kmol = 8 kmol + 9 kmol + 47 kmol

i.e.

$$1(114)\,kg + 12.5(32)\,kg + 47(28)\,kg = 8(44)\,kg + 9(18)\,kg + 47(28)\,kg$$

Therefore,

$$\text{mass of air/kg petrol} = \frac{12.5(32) + 47(28)}{114} = 15.05\,kg$$

11.3.4 Air–fuel ratio

The result from example 11.4 showed that petrol, assuming it to have the same composition as octane, requires 15.05 kg of air for the complete combustion of one kg of fuel. This proportion of air to fuel can be expressed more conveniently as the 'air–fuel ratio', defined as the mass of air to the mass of fuel during a combustion process:

$$A/F\,ratio = \frac{m_{air}}{m_{fuel}} \tag{11.5}$$

From example 11.4, the air–fuel ratio was found to be 15.05. However, this value represents the perfect ratio required for the **complete combustion** of the fuel. Such a perfect ratio is termed the 'stoichiometric ratio'. A stoichiometric ratio represents the ratio based upon perfect mixing of the fuel with the oxygen in the air to ensure complete combustion.

In actual combustion processes it is not possible to ensure perfect mixing so that more air is required than the stoichiometric amount. This increases the chance of complete combustion. For example, the air–fuel ratio for an oil fired boiler will be some 10 to 20% higher than the stoichiometric ratio to ensure that all the carbon in the fuel is burnt to carbon dioxide and that combustion is complete. Alternatively, the air–fuel ratio in the combustion chamber of a gas turbine engine is considerably greater than the stoichiometric ratio in order to control the temperature of the gases entering the turbine. The amount of excess air is generally expressed on a percentage basis:

$$\% \text{ excess air} = \left(\frac{\text{actual A/F}}{\text{stoichiometric A/F}} - 1 \right) \times 100 \tag{11.6}$$

For example, 100% excess air means that the actual A/F ratio is twice the stoichiometric ratio.

Example 11.5

Natural gas can be analysed assuming it to be 100% methane, CH_4. Calculate the stoichiometric air–fuel ratio on both a mass and volume basis.

Conceptual model

Analysis – the chemical equation for this combustion process can be written as

$$CH_4 + aO_2 + a3.76N_2 \rightarrow bCO_2 + cH_2O + a3.76N_2$$

Following the same procedure as used in the previous examples, the atom balance gives

$b = 1$
$c = 2$
$a = 2$

The chemical equation becomes

$$CH_4 + 2O_2 + 7.52N_2 \rightarrow CO_2 + 2H_2O + 7.52N_2$$
$$1\,kmol + 2\,kmol + 7.52\,kmol = 1\,kmol + 2\,kmol + 7.52\,kmol$$

i.e.

$$1\,(16)\,kg + 2\,(32)\,kg + 7.52\,(28)\,kg = 1\,(44)\,kg + 2\,(18)\,kg + 7.52\,(28)\,kg$$

Therefore,

$$A/F\ ratio\ (mass) = \frac{2(32) + 7.52(28)}{16} = 17.2$$

The air–fuel ratio, on a volume basis, can be found from the kmol balance of the chemical equation for complete combustion. Since a volume of gas contains a given number of molecules, it follows that a kmol is equivalent to a particular volume.

Rewriting the chemical equation,

$$CH_4 + 2O_2 + 7.52N_2 \rightarrow CO_2 + 2H_2O + 7.52N_2$$
$$1\,vol + 2\,vol + 7.52\,vol = 1\,vol + 2\,vol + 7.52\,vol$$

and

$$A/F\ ratio\ (volume) = \frac{2 + 7.52}{1} = 9.52$$

Note – the air–fuel ratio expressed on a volume basis is only applicable to a gaseous fuel such as natural gas. Unless specifically stated, an air–fuel ratio is generally quoted on a mass basis.

Example 11.6

Propane is a hydrocarbon fuel with a chemical formula C_3H_8. Calculate the actual air–fuel ratio for a combustion process with 25% excess air.

Conceptual model

Analysis – from equation (11.6), if the excess air is 25%,

actual A/F = 1.25 stoichiometric A/F

Assuming complete combustion the chemical equation for this process can be written as

$$C_3H_8 + 1.25(aO_2 + a3.76N_2) \rightarrow bCO_2 + cH_2O + 0.25aO_2 + 1.25a3.76N_2$$

The excess oxygen is assumed to be unchanged during the combustion process, and leaves as one of the products of combustion. The values of a, b and c can be evaluated by using the same procedure as for the previous examples.

carbon $3 = b$
hydrogen $8 = 2c$
oxygen $1.25(2a) = 2b + c + 0.25a$

It follows that

$b = 3$
$c = 4$
$a = 5$

The full chemical equation becomes

$$C_3H_8 + 6.25O_2 + 23.54N_2 \rightarrow 3CO_2 + 4H_2O + 1.25O_2 + 23.54N_2$$
1 kmol + 6.25 kmol + 23.54 kmol = 3 kmol + 4 kmol + 1.25 kmol + 23.54 kmol

i.e.

1 (44) kg + 6.25 (32) kg + 23.54 (28) kg = 3 (44) kg + 4 (18) kg + 1.25 (32) kg + 23.54 (28) kg

Therefore,

$$\text{A/F ratio} = \frac{6.25(32) + 23.54(28)}{44} = 19.52$$

Note – if the actual air–fuel ratio is 19.52 with 25% excess air, the stoichiometric air–fuel ratio is 19.52/1.25 = 15.62.

11.4   INTERNAL COMBUSTION ENGINES

The foregoing sections have discussed the chemical reactions that take place during combustion and have shown how the appropriate chemical equations can be derived in order to calculate the air–fuel ratio. Most liquid fuels that are used in internal combustion engines operate with a stoichiometric air–fuel ratio of around 15. The following table gives a list of three typical stoichiometric ratios for liquid fuels.

Fuel	Formula	Stoichiometric ratio
heptane	C_7H_{16}	15.10
octane	C_8H_{18}	15.05
kerosene	$C_{12}H_{24}$	14.71

Combustion can still take place if the actual air–fuel ratio is greater or less than stoichiometric, but there are limits. The limits tend to be about 40% greater, or less, than the stoichiometric ratio. Too great an amount of air quenches the flame and prevents it propagating amongst the reactants. Too little air provides insufficient oxygen to support combustion.

In the case of a spark ignition petrol engine, as typified by a car engine in which the fuel and air mix in a carburettor before entering the cylinder, the upper and lower limits of the operating air–fuel ratio are between 10 and 20. An air–fuel ratio greater than stoichiometric results in a 'weak' mixture. The air–fuel ratio will determine the power output and the fuel consumption of the engine.

As the air–fuel ratio varies, the point in the cycle at which combustion starts after ignition will vary. The velocity at which the flame travels through the reactants in the cylinder will also vary. These determine the shape of the cycle on the P–v diagram and, therefore, the net work output. If a petrol engine is run at constant rotational speed with constant throttle opening, the result of changing the air–fuel ratio is shown in Figure 11.4. The curve is plotted with net work output on the horizontal axis and thermal efficiency on the vertical axis.

Point 'b' represents the performance with a stoichiometric ratio. If the air–fuel ratio is increased the mixture becomes weaker and the net work is reduced, but the reduction in fuel flow results in an increased thermal efficiency to a maximum at 'a'. Reducing the air–fuel ratio results in the mixture becoming richer. The net work output increases to a maximum at 'c'. Although an engine operates over a wide range of rotational speeds, the performance can be plotted as a series of work–thermal efficiency curves with similar characteristics to figure 11.4. Points 'a' and 'c', either side of the stoichiometric ratio, represent the optimum operating air–fuel ratios for either minimum fuel consumption or maximum power output, respectively.

In the case of a gas turbine engine the actual air–fuel ratio might be of the order of 50, some 300% greater than the stoichiometric ratio. This is to reduce the temperature of the gases entering the turbine to a workable level. Such an air–fuel ratio would prevent combustion taking place if all the air was mixed with all the fuel in a single process. In practice, the air flow is divided and enters the combustion chamber in two streams, as shown in Figure 11.5.

The air flow through the combustion chamber of a gas turbine engine is divided into the 'primary' air flow and the 'secondary' air flow. The primary air enters through front of the chamber and is mixed with the fuel at near stoichiometric ratios to support combustion. The rest of the air, in fact the greater proportion, flows around the outside of the chamber before, gradually entering

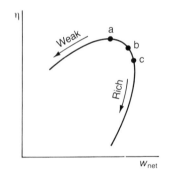

Figure 11.4 Work–thermal efficiency curve.

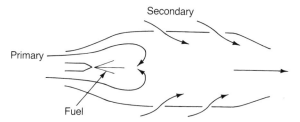

Figure 11.5 Combustion chamber of a gas turbine engine.

to mix with the products of combustion. This secondary air cools the gases to the required turbine inlet temperature.

11.5 STEADY FLOW COMBUSTION

The whole purpose of combustion is to bring about a release of thermal energy. The thermal energy released can be found by analysing the process within a combustion chamber using the steady flow energy equation.

Figure 11.6 shows a steady flow combustion chamber in which a combustion process takes place at constant pressure. The combustion process is assumed to take place with air as the oxidant. One kmol of fuel enters the combustion chamber with the appropriate amount of air. These form the reactants. Taking the air to comprise oxygen and nitrogen only, the reactants can be considered to consist of three streams. These three streams being the fuel, oxygen and nitrogen as three separate substances within the chemical reaction.

Similarly, the products can be considered to consist of several streams, each stream representing a separate substance.

Applying the steady flow energy equation, (5.5), to a combustion chamber it has been shown that when the velocities of the streams are ignored, the heat transferred is given by equation (5.6):

$$q = h_2 - h_1$$

However, these quantities are based upon a unit mass, whereas combustion processes are more conveniently expressed on a kmol basis. The steady flow energy equation for a combustion process can, therefore, be expressed in the form

$$\bar{q} = \bar{h}_2 - \bar{h}_1 \tag{11.7}$$

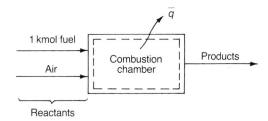

Figure 11.6 Steady flow combustion model.

where $\bar{q}$ represents the heat transferred during the combustion process per kmol and $\bar{h}$ is the enthalpy expressed in kJ/kmol.

Equation (11.7) is based on the assumption that 1 kmol of the reactants enters the combustion chamber and 1 kmol of the products leaves. In practice, this may be impossible to achieve. Whereas a combustion process must maintain a mass balance it is not necessary to achieve a kmol balance. In order to apply equation (11.7) to a combustion process the steady flow equation has to be modified to take into account the different proportions of the reactants and the products.

It can be assumed that the analysis of the combustion process is based on 1 kmol of the fuel entering the combustion chamber. This then represents the starting point for the analysis. To ensure combustion, air or oxygen must also enter with the fuel as the reactants. So the total enthalpy of the reactants can be expressed as

$$H_R = \bar{h}_{fuel} + n_{O_2}\,\bar{h}_{O_2} + n_{N_2}\,\bar{h}_{N_2}$$

where the subscript 'R' for the total enthalpy H refers to the reactants and n denotes the number of kmol.

Similarly, the total enthalpy of the products can be denoted by the symbol H_P.

The difference between the total enthalpy of the products and the total enthalpy of the reactants is the quantity of the heat transferred. Since this is based on the combustion of 1 kmol of fuel, it can still be expressed as $\bar{q}$ and the steady flow energy equation for combustion becomes

$$\bar{q} = H_P - H_R \tag{11.8}$$

This is true for **all** steady flow combustion processes **irrespective** of the number of streams of the reactants and products.

Enthalpy is not an absolute property, its value must always be related to a particular datum. For combustion processes the values of enthalpy are taken to be zero at a temperature of 25°C (298 K). In fact, for combustion, the standard reference conditions are taken as a temperature of 25°C when the pressure in the combustion chamber is one standard atmosphere, 101.325 kPa.

11.6   THERMAL ENERGY OF COMBUSTION

Combustion is a chemical reaction in which the bonds between the atoms within a molecule of fuel are broken to allow new compounds to be formed. The formation of a new compound is associated with a change of energy. If the formation of a new compound takes place at constant pressure, the change of energy can be expressed as a change of enthalpy and analysed using the steady flow energy equation defined in equation (11.8).

11.6.1   Enthalpy of formation

The enthalpy of formation $\bar{h}_f$ is the change of energy associated with the formation of a particular compound from its constituent elements under standard reference conditions. All elements that have a stable form under standard

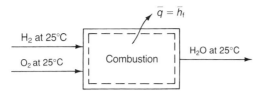

Figure 11.7 Enthalpy of formation of water.

conditions have an enthalpy value of zero. Oxygen, O_2, nitrogen, N_2, and hydrogen, H_2, represent chemically stable forms of these elements at the standard reference temperature of 25°C. Therefore, they all have enthalpy values of $\bar{h}_f = 0$ at this temperature.

If such stable elements are brought together within a chemical reaction to form a new compound, there will be a change of energy which represents the enthalpy of formation for that compound.

Figure 11.7 shows a constant pressure combustion process in which hydrogen and oxygen are combined to form water. The hydrogen and oxygen enter the combustion chamber as reactants at the standard reference temperature of 25°C. The water leaves as the product at the same temperature. For this reaction, the chemical equation is defined by equation (11.2):

$$H_2 \quad + \tfrac{1}{2}O_2 \quad \rightarrow H_2O$$
$$1 \text{ kmol} + \tfrac{1}{2} \text{ kmol} = 1 \text{ kmol}$$

At a temperature of 25°C the water will be in a liquid state and the heat transferred during the process will be 285 770 kJ/kmol of H_2O. Since the heat transfer is from the combustion chamber to the surroundings, it follows that

$$\bar{q} = -285\ 770 \text{ kJ/kmol of } H_2O$$

As the two reactants, H_2 and O_2, are both stable elements under standard conditions, the enthalpy of both will be zero and $H_R = 0$. It follows that the heat transferred during the process must equal the enthalpy of formation, so that the enthalpy of formation $\bar{h}_f$ for water in a liquid state is –285 770 kJ/kmol.

For most combustion processes water is only one constituent of the products and the partial pressure will be such that the water is a vapour. The enthalpy of formation for water in a vapour state is –241 820 kJ/kmol. A table of typical values of the enthalpy of formation for a range of compounds is given in Appendix C1.

11.6.2 Enthalpy of combustion

The enthalpy of combustion $\bar{h}_c$ is defined as the amount of thermal energy released during a constant pressure combustion process when 1 kmol of fuel is completely burnt under standard conditions.

A combustion process involves the fuel and air, or oxygen, as the reactants. The products will include several compounds, each of which will release thermal energy according to its own enthalpy of formation and the quantity in the products.

For example, a hydrocarbon fuel being completely burnt with a stoichiometric ratio of air will have carbon dioxide, water vapour and nitrogen as the constituents of the products. If the products are at a standard temperature of 25°C, N_2 is a stable element and will have zero enthalpy of formation value. Therefore, the total enthalpy of the products will be

$$H_P = (n\overline{h}_f)_{CO_2} + (n\overline{h}_f)_{H_2O}$$

Similarly, the reactants will consist of 1 kmol of fuel together with the required amount at air. Air consists of oxygen and nitrogen, both of which are stable elements having zero enthalpy of formation values at 25°C. Therefore, the total enthalpy of the reactants will be

$$H_R = (\overline{h}_f)_{fuel}$$

The heat transferred during such a combustion process will be the enthalpy of combustion. Applying the steady flow energy equation, (11.8),

$$\overline{h}_c = \overline{q} = H_P - H_R$$

therefore,

$$\overline{h}_c = (n\overline{h}_f)_{CO_2} + (n\overline{h}_f)_{H_2O} - (\overline{h}_f)_{fuel} \tag{11.9}$$

where the value of $\overline{h}_c$ is based on 1 kmol of fuel.

Example 11.7

Calculate the enthalpy of combustion at a standard temperature of 25°C for the complete combustion of liquid heptane (C_7H_{16}).

Conceptual model

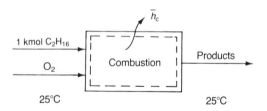

Analysis – since nitrogen has $\overline{h}_f = 0$, it will play no part in the analysis and the combustion process can be assumed to take place with oxygen alone. The chemical equation for this combustion process can be written as

$$C_7H_{16} + aO_2 \rightarrow bCO_2 + cH_2O$$

Following the same procedure as used in the previous examples, the atom balance gives

$$b = 7$$
$$c = 8$$
$$a = 11$$

The chemical equation becomes

$$C_7H_{16} + 11O_2 \rightarrow 7CO_2 + 8H_2O$$
$$1\,kmol + 11\,kmol = 7\,kmol + 8\,kmol$$

Taking values of enthalpy of formation from Appendix C1,

$$(\bar{h}_f)_{C_7H_{16}} = -224\,390\ kJ/kmol$$
$$(\bar{h}_f)_{CO_2} = -393\,520\ kJ/kmol$$
$$(\bar{h}_f)_{H_2O} = -241\,820\ kJ/kmol$$

Substituting in equation (11.9)

$$
\begin{aligned}
\bar{h}_c &= (n\bar{h}_f)_{CO_2} + (n\bar{h}_f)_{H_2O} - (\bar{h}_f)_{fuel}\\
&= 7(-393520) + 8(-241820) - (-224390)\\
&= -4\,464\,810\ kJ/kmol
\end{aligned}
$$

Note – this value is base on the assumption that the water in the products is in the form of a vapour.

11.6.3 Energy content of fuels

The energy content of a fuel is sometimes referred to as its 'calorific value'. Since this value is defined as the quantity of thermal energy released during complete combustion, it must be equal to the enthalpy of combustion, discussed in the last section. The only real difference is that enthalpy of combustion is quoted in kJ/kmol, whereas the calorific value of a solid or liquid fuel is quoted in kJ/kg.

Taking the enthalpy of combustion value of $-4\,464\,810$ kJ/kmol from example 11.6, this represents the thermal energy released per kmol of fuel. Dividing by the mass in 1 kmol of C_7H_{16}, i.e. 100 kg, gives a thermal energy value of 44 648 kJ/kg. This will be seen to be very close to the energy content of petrol given in section 11.2.

Calorific values are quoted as either a higher calorific value or a lower calorific value. The difference between the two is given by the state of the water in the products of combustion. If the water is in a liquid state the resulting thermal energy released is defined by the higher calorific value. If the water is in a vapour state there is an increase of the enthalpy of the products which **reduces** the thermal energy that can be released, resulting in a **lower** calorific value.

This is also true for the enthalpy of combustion. The value of $-4\,464\,810$ kJ/kmol, calculated in example 11.6, was based on the assumption that the water in the products was in a vapour state. If the water had been assumed to be in a liquid state the enthalpy of combustion would have been $-4\,816\,410$ kJ/kmol, an increase of some 8%. In practice, water is just one constituent of the products of combustion and its partial pressure will be sufficiently low so as to ensure that it is in a vapour state.

Example 11.8

Calculate the lower calorific value at a standard temperature of 25°C for the complete combustion of propane (C_3H_8).

Conceptual model

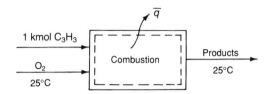

Analysis – the chemical equation for this combustion process can be written as

$$C_3H_8 + aO_2 \rightarrow bCO_2 + cH_2O$$

Following the same procedure as used in the previous examples, the atom balance gives

$b = 3$
$c = 4$
$a = 5$

The chemical equation becomes

$$C_3H_8 + 5O_2 \rightarrow 3CO_2 + 4H_2O$$

Taking values of enthalpy of formation from Appendix C1,

$(\bar{h}_f)_{C_3H_8}$ $= -103\ 840$ kJ/kmol
$(\bar{h}_f)_{CO_2}$ $= -393\ 520$ kJ/kmol
$(\bar{h}_f)_{H_2O}$ $= -241\ 820$ kJ/kmol

Substituting in equation (11.9)

$$
\begin{aligned}
\bar{h}_c &= (n\bar{h}_f)_{CO_2} + (n\bar{h}_f)_{H_2O} - (\bar{h}_f)_{fuel} \\
&= 3(-393\ 520) + 4(-241\ 820) - (-103\ 840) \\
&= -2\ 044\ 000 \text{ kJ/kmol}
\end{aligned}
$$

Since the molecular weight of propane is 44 the mass in 1 kmol is 44 kg, and

calorific value = 2 044 000/44 = 46 454.5 kJ/kg

Note – this is the lower calorific value as the enthalpy of formation for water is based upon the assumption that it is in a vapour form.

11.7   ADIABATIC COMBUSTION

The above title might seem somewhat strange because the whole purpose of combustion is to release thermal energy, whereas an adiabatic process is one in

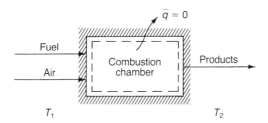

Figure 11.8 Adiabatic combustion process.

which there is no heat transfer. However, an adiabatic combustion process is a realistic one because the aim is generally not to increase the temperature of the surroundings but to increase the temperature of the products.

An adiabatic combustion process is shown in Figure 11.8. The combustion chamber is assumed to be insulated so that heat transfer between the surface of the chamber and the surroundings is zero. This is typical of a combustion chamber in a gas turbine engine in which the purpose of combustion is to provide a temperature increase to the gas flow. As a result, the temperature of the products T_2 will not only be higher than the reference temperature of 25°C but also higher than the inlet temperature of T_1.

The increase in temperature can be found using the steady flow energy equation for combustion, equation (11.8). Alternatively, the temperature of the products can be controlled by the amount of air and the same equation can be used to find the required air–fuel ratio to give a particular temperature T_2. Since the external heat transfer is zero, equation (11.8) can be written as

$$0 = H_P - H_R$$

and this is true for any number of streams entering or leaving the combustion chamber.

Taking one compound as a stream at a temperature T, the total enthalpy will be

$$H_{\text{stream}} = n\overline{h}$$

where $\overline{h}$ is the enthalpy of the compound at temperature T.

Apart from the fuel, all the streams can be assumed to be gases. Enthalpy values for the gases found in combustion processes are quoted in Appendix C2. At the reference temperature of 25°C the enthalpy of each gas is equal to its enthalpy of formation. At temperatures above 25°C the enthalpy of each gas increases due to the increase in temperature, $(T - 298)$. However, this increase in enthalpy is not linear as C_P also varies with temperature. The values of enthalpy quoted in Appendix C2 are for a working range of temperatures, 298 to 2400 K.

Combining the total enthalpy for each stream in the products give H_P, while combining the total enthalpy for each stream in the reactants gives H_R. These values can be substituted in equation (11.8). The following example illustrates how this equation can be applied.

Example 11.9

A gas turbine engine operates with butane (C_4H_{10}) as the fuel. The fuel enters the combustion chamber at 25°C and is mixed with excess air entering at 600 K.

Assuming the combustion process to be adiabatic, calculate the air–fuel ratio required to maintain the products at a temperature of 1200 K.

Conceptual model

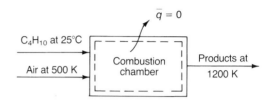

Analysis – from example 11.1 the chemical equation for the combustion of butane with oxygen is

$$C_4H_{10} + 6.5O_2 \rightarrow 4CO_2 + 5H_2O$$

Therefore, the equation for complete combustion with air is

$$C_4H_{10} + 6.5(O_2 + 3.76N_2) \rightarrow 4CO_2 + 5H_2O + 24.44N_2$$

The actual combustion process will require excess air to reduce the temperature of the products to 1200 K. Defining x as the number of kmol of air,

$$C_4H_{10} + xO_2 + 3.76xN_2 \rightarrow 4CO_2 + 5H_2O + (x-6.5)O_2 + 3.76xN_2$$
$$1\text{ kmol} + x\text{ kmol} + 3.76x\text{ kmol} = 4\text{ kmol} + 5\text{ kmol} + (x-6.5)\text{ kmol} + 3.76x\text{ kmol}$$

298 K   600 K   1200 K

The enthalpies for the reactants and the products can be found using Appendices C1 and C2. For the reactants

$$H_R = -126\ 140 + x(9249) + 3.76x(8895)$$
$$= -126\ 140 + 42\ 694.2x\text{ kJ}$$

For the products

$$H_P = 4(-349\ 041) + 5(-207\ 323) + x(29\ 758) - 6.5(29\ 758) + 3.76x(28\ 110)$$
$$= -2\ 626\ 206 + 135\ 451.6x\text{ kJ}$$

For adiabatic combustion

$$0 = H_P - H_R$$
$$= -2\ 626\ 206 + 135\ 451.6x + 126\ 140 - 42\ 694.2x$$
$$= -2\ 500\ 066 + 92\ 757.4x$$

Therefore

$$x = \frac{2\,500\,066}{92\,757.4} = 26.95$$

To find the air–fuel ratio

1 kmol C_4H_{10} = 58 kg
26.95 kmol $(O_2 + 3.76N_2)$ = 26.95(32) + 101.33(28) kg

Therefore,

$$\text{A/F ratio} = \frac{26.95(32) + 101.33(28)}{58} = 63.8$$

Note – this air–fuel ratio is typical of gas turbine engines and confirms the assumption made when analysing the open cycle gas turbine engine, that the mass of fuel is negligible compared to the air.

Example 11.10

Octane (C_8H_{18}) enters a combustion chamber at 25°C and is mixed with just the correct amount of air for complete combustion. If the air enters at 25°C, estimate the temperature of the products if the combustion process can be assumed to be adiabatic.

Conceptual model

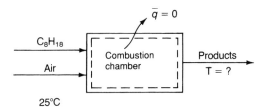

Analysis – the solution to this problem requires a 'trial and error' approach, as outlined below. From example 11.4, the chemical equation for the complete combustion of octane is

$$C_8H_{18} + 12.5O_2 + 47N_2 \rightarrow 8CO_2 + 9H_2O + 47N_2$$
$$\underbrace{1\,\text{kmol} + 12.5\,\text{kmol} + 47\,\text{kmol}}_{25°C} = \underbrace{8\,\text{kmol} + 9\,\text{kmol} + 47\,\text{kmol}}_{T}$$

The enthalpies for the reactants and products can be found using Appendices C1 and C2. For the reactants

$$\begin{aligned} H_R &= -249\,950 + 12.5(0) + 47(0) \\ &= -249\,950 \text{ kJ} \end{aligned}$$

Assuming adiabatic combustion

$$0 = H_P - H_R$$

therefore,

$$H_P = H_R = -249\ 950\ \text{kJ}$$

The solution requires a value of T that is appropriate to this value of H_P. Assuming $T = 2000$ K,

$$H_P = 8(-302\ 078) + 9(-169\ 065) + 47(56\ 156)$$
$$= -1\ 298\ 877\ \text{kJ}$$

This value of H_P is too low, indicating that the assumed value of T is too low. assuming $T = 2400$ K

$$H_P = 8(-277\ 737) + 9(-148\ 139) + 47(70\ 661)$$
$$= -234\ 080\ \text{kJ}$$

This value of H_P is slightly high, indicating that the assumed value of T is too high. By interpolating between these values the final temperature is near 2395 K.

Note – the temperature of the products estimated above is sometimes referred to as the adiabatic flame temperature.

SUMMARY

In this chapter the chemical reactions and energy changes that take place during combustion processes have been discussed.

The key terms that have been introduced are:

hydrocarbon fuels
oxidant
reactants
products of combustion
ignition temperature
liquid fuel rocket engine
air–fuel ratio
stoichiometric ratio
percentage excess air
weak mixture
rich mixture
steady flow combustion
enthalpy of formation
enthalpy of combustion
calorific value
adiabatic combustion

Key equations that have been introduced.

For combustion with oxygen:

$$C + O_2 \rightarrow CO_2 \qquad (11.1)$$
$$H_2 + \tfrac{1}{2}O_2 \rightarrow H_2O \qquad (11.2)$$

For complete combustion with air:

$$C + O_2 + 3.76N_2 \rightarrow CO_2 + 3.76N_2 \qquad (11.3)$$

$$H_2 + \frac{1}{2}O_2 + \frac{3.76}{2}N_2 \rightarrow H_2O + \frac{3.76}{2}N_2 \qquad (11.4)$$

and the air–fuel ratio is defined as

$$A/F \ \mathrm{ratio} = \frac{m_{air}}{m_{fuel}} \qquad (11.5)$$

Excess air given as a percentage:

$$\% \ \mathrm{excess \ air} = \left(\frac{\mathrm{actual \ A/F}}{\mathrm{stoichiometric \ A/F}} - 1 \right) \times 100 \qquad (11.6)$$

The steady flow energy equation for combustion:

$$\bar{q} = H_P - H_R \qquad (11.8)$$

Enthalpy of combustion for a hydrocarbon fuel:

$$\bar{h} = (n\bar{h}_f)_{CO_2} + (n\bar{h}_f)_{H_2O} - (\bar{h}_f)_{fuel} \qquad (11.9)$$

PROBLEMS

1. Find the stoichiometric air–fuel ratio for the complete combustion of benzene (C_6H_6).

2. Pentane (C_5H_{12}) is burnt with 20% excess air. Calculate the air–fuel ratio assuming complete combustion.

3. In a rocket engine, kerosine ($C_{12}H_{24}$) is burnt with hydrogen peroxide (H_2O_2). Assuming the products of combustion to be carbon dioxide and water vapour, calculate the mass of hydrogen peroxide required for each kg of fuel.

4. Petrol consists of 80% C_8H_{18} and 20% C_7H_{16} by mass. If it is burnt with an air–fuel ratio of 18, calculate the percentage excess air.

5. The complete combustion of 1 kmol of a hydrocarbon fuel results in the following products of combustion: $7CO_2$, $8H_2O$, $2O_2$ and $48.9N_2$ on a kmol basis.
   Determine the chemical composition of the fuel and the percentage excess air.

6. Determine the enthalpy of combustion of gaseous ethane (C_2H_6) at 25°C, assuming that the water in the products of combustion is in a vapour state.

7. Find the lower calorific value for octane (C_8H_{18}) undergoing complete combustion at 25°C.

8. Butane (C_4H_{10}) enters an insulated combustion chamber with excess air. If the combustion is complete and both the ethane and air enter at 25°C, calculate the air–fuel ratio required to maintain the products at 1800 K.

9. A gas turbine engine burns natural gas. The gas enters the combustion chamber at 25°C with air at 800 K. Calculate the percentage excess air required to ensure a turbine inlet temperature of 1400 K. Assume natural gas to comprise only methane (CH_4).

Heat transfer $\boxed{12}$

12.1 AIMS

- To introduce the modes of heat transfer:

  convection
  conduction
  radiation

- To define the overall heat transfer coefficient for combined modes of heat transfer.
- To evaluate the overall heat transfer coefficient using a thermal resistance analogy.
- To introduce the mean temperature difference for a heat exchanger and define:
  arithmetic mean temperature difference;
  log mean temperature difference.
- To compare the use of the arithmetic and log mean temperature differences.
- To define the two main types of heat exchanger:

  in-line
  cross-flow

- To discuss the use of counter-flow or parallel-flow, shell and tube heat exchangers.
- To discuss the use of finned tubes for cross-flow heat exchangers with air flow across the tubes.

12.2 MODES OF HEAT TRANSFER

As explained in Chapter 3, heat is a form of energy that is transferred across the boundary of a thermodynamic system as a result of a temperature difference. The greater the temperature difference the more rapidly will the heat be transferred. Conversely, the lower the temperature difference, the slower will be the rate at which heat is transferred. When discussing the modes of heat transfer it is the rate of heat transfer Q that defines the characteristics rather than the quantity of heat.

There are three distinct modes of heat transfer, convection, conduction and radiation. Although two, or even all three, modes of heat transfer may be combined in any particular thermodynamic situation, the three are quite different and will be introduced separately.

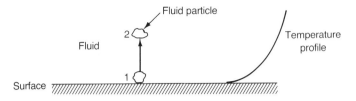

Figure 12.1 Mechanism of convection.

12.2.1 Convection

Convection is a mode of heat transfer that takes place as a result of motion within a fluid. Consider a particle of fluid, as shown in Figure 12.1, moving from position '1' to position '2' within a region of the fluid adjacent to a surface.

If the fluid, shown in Figure 12.1, starts at a constant temperature and the surface is suddenly increased in temperature to above that of the fluid, there will be convective heat transfer from the surface to the fluid as a result of the temperature difference. The particle at '1' starts in contact with surface and its temperature will rise as a result of that contact. If the particle then moves to position '2', it will have a higher temperature than the fluid around it. As a result, the temperature of the particle will fall as heat is transferred to the fluid around it, until the particle and the surrounding fluid reach thermal equilibrium.

Within this model, heat is transferred from the surface to the fluid in the region of position '2' by means of movement of one fluid particle. In real situations, all the particles within the fluid will be in motion and there will be a gradual drop in temperature from the surface through the subsequent layers of the fluid, as defined by the temperature profile shown in Figure 12.1. Under these conditions the temperature difference causing the heat transfer can be defined as

ΔT = surface temperature – mean fluid temperature

Using this definition of the temperature difference, the rate of heat transfer due to convection can be evaluated using Newton's law of cooling:

$$Q = h_c A \, \Delta T \tag{12.1}$$

where A is the heat transfer surface area and h_c is the coefficient of heat transfer from the surface to the fluid, referred to as the 'convective heat transfer coefficient'.

The units of the convective heat transfer coefficient can be determined from the units of the other variables:

$$Q = h_c A \, \Delta T$$
$$W = (h_c) m^2 \, K$$

so the units of h_c are $W/m^2 \, K$.

The relationship given in equation (12.1) is also true for the situation where a surface is being heated due to the fluid having a higher temperature than the

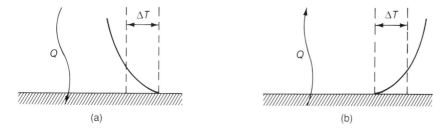

Figure 12.2 Convective heat transfer situations.

surface. However, in this case the direction of heat transfer is from the fluid to the surface and the temperature difference will now be

ΔT = mean fluid temperature − surface temperature

The relative temperatures of the surface and fluid determine the direction of heat transfer and the rate at which heat transfer takes place. Figure 12.2 illustrates the two alternative convective situations of heating and cooling.

Figure 12.2(a) defines the situation in which a surface is being 'heated' by a fluid at a higher temperature, so that there is a temperature drop from the fluid to the surface. Figure 12.2(b) defines the situation in whcih the surface is being cooled by a fluid at a lower temperature, so that there is a temperature drop from the surface to the fluid.

As given in equation (12.1), the rate of heat transfer is not only determined by the temperature difference but also by the convective heat transfer coefficient h_c. This is not a constant but varies quite widely depending on the properties of the fluid and the behaviour of the flow.

The simple model defined in Figure 12.1 indicates that the value of h_c must depend on the thermal capacity of the fluid particle considered, i.e. mC_P for the particle. In other words the higher the density and C_P of the fluid the better the convective heat transfer.

Two common heat transfer fluids are air and water, due to their widespread availability. Water is approximately 800 times more dense than air and also has a higher value of C_P. If the argument given above is valid then water has a higher thermal capacity than air and should have a better convective heat transfer performance. This is borne out in practice because typical values of convective heat transfer coefficients are as follows:

Fluid	h_c (W/m² K)
water	500–10000
air	5–100

The variation in the values reflects the variation in the behaviour of the flow, particularly the flow velocity, with the higher values of h_c resulting from higher flow velocities over the surface.

12.2.2 Conduction

If a fluid could be kept stationary there would be no convection taking place. However, it would still be possible to transfer heat by means of conduction.

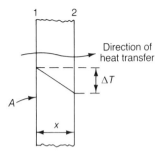

Figure 12.3 Conduction in one direction.

Conduction depends on the transfer of energy from one molecule to another within the heat transfer medium and, in this sense, thermal conduction is analogous to electrical conduction.

Conduction can occur within both solids and fluids. The rate of heat transfer depends on a physical property of the particular solid or fluid, termed its thermal conductivity k, and the temperature gradient across the medium. The thermal conductivity is defined as the measure of the rate of heat transfer across a unit width of material, for a unit cross-sectional area and for a unit difference in temperature.

Figure 12.3 defines a conduction situation in which heat transfer is taking place across a solid wall. In more complex situations it is possible to have conduction taking place simultaneously in three directions, but for the present discussion it is sufficient to consider conduction in just one direction.

From the definition of thermal conductivity k it can be shown that the rate of heat transfer is given by the relationship

$$Q = \frac{kA\,\Delta T}{x} \tag{12.2}$$

where ΔT is the temperature difference $T_1 - T_2$, defined by the temperatures on the either side of the wall.

The units of thermal conductivity can be determined from the units of the other variables:

$$Q = k\,A\,\Delta T/\text{x}$$
$$W = (k)\text{m}^2\,\text{K/m}$$

so the units of k are W/m²K/m, expressed as W/mK.

Since thermal conductivity is analogous to electrical conductivity, it would follow that metals having good electrical conductivity should have good thermal conductivity. Similarly, electrical insulators should have low thermal conductivity. This is borne out in Table 12.1.

Table 12.1 Thermal conductivities for some common materials

Material	Thermal conductivity (W/mK)
Metals	
aluminium	210
copper	360
steel	44
Building materials	
brick	0.7
concrete	0.8
wood	0.2
Thermal insulation	
asbestos	0.1
foam plastic	0.04
glass fibre	0.05

Example 12.1

A furnace has a refractory wall of 0.1 m thickness and thermal conductivity of 1 W/mK. The convective heat transfer coefficient between the wall and the outside air is 10 W/m²K. If the inner surface of the furnace is at 800°C and the outside air is at 20°C, find the outside temperature of the wall.

Conceptual model

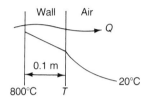

Analysis — heat transfer across the wall is by conduction. From equation (12.2)

$$Q = \frac{kA\,\Delta T}{x} = \frac{1\,(800 - T)}{0.1}A$$

therefore,

$$\frac{Q}{A} = 8000 - 10T$$

Heat transfer from the wall to the air is by convection.
From equation (12.1)

$$Q = h_c\,A\,\Delta T = 10\,(T - 20)A$$

therefore,

$$\frac{Q}{A} = 10T - 200$$

Assuming Q/A to be the same for both modes of heat transfer,

$$8000 - 10T = 10T - 200$$

and

$$T = \frac{8200}{20} = 410°C$$

12.2.3 Radiation

The third mode of heat transfer, radiation, does not depend on any medium for its transmission. In fact, it takes place most freely when there is a perfect vacuum between the emitter and the receiver of such energy. This is proved daily by the transfer of energy from the sun to the earth across the intervening space.

Radiation is a form of electromagnetic energy transmission and takes place between all matter providing that it is at a temperature above absolute zero. Infra-red, visible light and ultra-violet radiation form just part of the overall electromagnetic spectrum. Radiation is energy emitted by the electrons vibrating in the molecules at the surface of a body. The amount of energy that can be transferred depends on the absolute temperature of the body and the radiant properties of the surface.

A body that has a surface that will absorb all the radiant energy it receives is an ideal radiator, termed a 'black body'. Such a body will not only **absorb** radiation at a maximum level but will also **emit** radiation at a maximum level. However, in practice, bodies do not have the surface characteristics of a black body and will always absorb, or emit, radiant energy at a lower level than a black body.

It is possible to define how much of the radiant energy will be absorbed, or emitted, by a particular surface by the use of a correction factor, known as the 'emissivity' and given the symbol $\in$. The emissivity of a surface is the measure of the actual amount of radiant energy that can be absorbed, compared to a black body. Similarly, the emissivity defines the radiant energy emitted from a surface compared to a black body. A black body would, therefore, by definition, have an emissivity $\in$ of 1. Typical values of emissivity for some common materials found in practice are:

Material	Emissivity ($\in$)
aluminium, polished	0.05
oxidized	0.10
steel, polished	0.30
oxidized	0.70
building brick	0.85
glass	0.94
paint, white	0.85
black matt	0.97

From the above list it will be seen that the value of emissivity is influenced more by the nature of the surface, than its colour. The practice of wearing white clothes in preference to dark clothes in order to keep cool on a hot summer's day is not necessarily valid. The amount of radiant energy absorbed is more a function of the **texture** of the clothes rather than the colour.

Using the emissivity, the rate of heat transfer due to radiation can be evaluated using an equation derived by Stefan and Boltzmann in the latter part of the 19th century (Simonson, 1988). Taking a surface at a temperature T_1 and an emissivity of $\in$, that is radiating energy to a large surrounding space at a lower constant temperature of T_2, the heat transferred from the surface is given by the relationship

$$Q = \sigma \in A \, (T_1^4 - T_2^4) \tag{12.3}$$

where σ is the Stefan–Boltzmann constant, 5.67×10^{-8} W/m^2 K^4 and A is the surface area.

For example, equation 12.3 would be true of the earth radiating out into space. Assuming the atmosphere to be transparent to radiation there would be direct transfer of heat from the surface of the earth out into space. Nitrogen and oxygen, the major constituents of the atmosphere are, for all practical purposes, transparent to radiation. However, if there is a cloud layer above the surface the radiant energy loss is reduced. This is why a clear night sky will cause a heavy frost during the winter, while a cloud layer could help maintain temperatures above freezing.

As well as clouds, carbon dioxide and water vapour are present as minor constituents of the air and do absorb radiant energy. The actual absorption depends on the partial pressure of these gases as constituents in a mixture of gases and the free length that the radiation travels through the gas. To given an indication of the influence of such gases, a partial pressure of 10 kPA and a free length of 0.5 m within a gas mixture at standard atmospheric temperature and pressure, would result in effective emissivity values of 0.036 for carbon dioxide and 0.042 for water vapour.

Of course, the partial pressures of these constituents would generally be much lower than 10 kPa but the free length would be much greater than 0.5 m. Nevertheless, the value of emissivity quoted for carbon dioxide above indicates why it is considered as a 'greenhouse' gas. Its presence in the atmosphere ensures that not only is more solar radiation absorbed but that the radiant loss at night is reduced. As the amount of carbon dioxide in the atmosphere is known to be increasing, it is clearly going to have an effect on the future climate of the world.

12.3 OVERALL HEAT TRANSFER COEFFICIENT

In reality, the three modes of heat transfer do not occur as isolated events. In fact, for most practical situations, heat transfer relies on two, or even all three, modes occurring together. For such situations, it is inconvenient to analyse each mode separately. Therefore, it is useful to derive an overall heat transfer coefficient that will combine the effect of each mode within a general situation.

12.3.1 Combined convection and conduction

A central heating unit in which hot water flows through a heat exchanger and so heats the air in a room, is termed a 'radiator'. Similarly, an air-cooled heat exchanger, necessary to cool the hot water from a car engine is also termed a 'radiator'. With the widespread use of such a term it would be imagined that radiation is the most important mode of heat transfer. However, this is **not** the case. Both types of radiator described above rely on a combination of convection and conduction, with radiation playing only a small part in the operation.

In the case of a central-heating radiator, the heat transfer processes involve convection from the hot water to the inner surface of the heat exchanger, conduction across the metal wall and convection from the outer surface to the surrounding air. These combined modes of convection and conduction are shown in Figure 12.4.

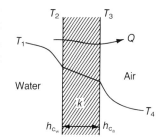

Figure 12.4 Combined convection and conduction.

Assuming the convective heat transfer coefficients for water and air are h_{c_w} and h_{c_a} respectively, and the thermal conductivity of the wall is k, the situation defined in Figure 12.4 can be analysed as follows.

Water side — from equation (12.1)

$$Q = h_{c_w} A (T_1 - T_2)$$

therefore,

$$T_1 - T_2 = \frac{Q}{A h_{c_w}}$$

Across the wall – from equation (12.2)

$$Q = \frac{kA}{x} (T_2 - T_3)$$

therefore,

$$T_2 - T_3 = \frac{Qx}{Ak}$$

Air-side — from equation (12.1)

$$Q = h_{c_a} A (T_3 - T_4)$$

therefore,

$$T_3 - T_4 = \frac{Q}{A h_{c_a}}$$

Assuming that the rate of heat transfer Q is the same across the combined situation, the temperature differences can be summed to give

$$(T_1 - T_2) + (T_2 - T_3) + (T_3 - T_4) = \frac{Q}{A h_{c_w}} + \frac{Qx}{Ak} + \frac{Q}{A h_{c_a}}$$

resulting in

$$T_1 - T_4 = \frac{Q}{A} \left(\frac{1}{h_{c_w}} + \frac{x}{k} + \frac{1}{h_{c_a}} \right)$$

(12.4)

The temperature difference $(T_1 - T_4)$ represents the difference for the overall situation and can be expressed as ΔT. Similarly, the surface heat transfer coefficients and the thermal conduction across the wall can be combined into an overall heat transfer coefficient U, defined by

$$\frac{1}{U} = \frac{1}{h_{c_w}} + \frac{x}{k} + \frac{1}{h_{c_a}}$$

(12.5)

Combining equations (12.4) and (12.5) gives a general relationship:

$$Q = U A \Delta T$$

(12.6)

This relationship is based upon an **overall heat transfer coefficient** which includes convection on both sides of a plane wall and thermal conduction across the wall. By using similar models to that given in Figure 12.4, the overall heat transfer coefficient can be derived for any combination of convection and conduction.

Example 12.2

A central-heating radiator is constructed from sheet steel, 2 mm thick. If the surface heat transfer coefficients are 1000 W/m²K and 10 W/m²K, and the temperatures are 80°C and 20°C, on the water- and air-sides respectively, calculate the required surface area for a rate of heat transfer of 2 kW. Assume the thermal conductivity for steel to be 40 W/mK.

Conceptual model — Figure 12.4.

Analysis — the overall temperature difference is

$$\Delta T = 80 - 20 = 60 \text{ K}$$

The overall heat transfer coefficient is defined by equation (12.5):

$$\frac{1}{U} = \frac{1}{h_{c_w}} + \frac{x}{k} + \frac{1}{h_{c_a}}$$

$$= \frac{1}{1000} + \frac{2}{1000 \times 40} + \frac{1}{10}$$

$$= 0.001 + 0.00005 + 0.1 = 0.10105 \text{ m}^2\text{K/W}$$

Therefore,

$$U = 9.9 \text{ W/m}^2\text{K}$$

Substituting in equation (12.6)

$$Q = UA\,\Delta T$$
$$2000 = 9.9 \times A \times 60$$
$$A = 2000/9.9 \times 60 = 3.37 \text{ m}^2$$

12.3.2 Thermal resistance

The geometry described above represents quite a simple heat transfer situation. Not all heat exchangers employ water or air as the working fluids, there are a range of other fluids·that can be used although water and air tend to be used most widely. Similarly, not all heat transfer situations involve a plane wall made of a single material. Many involve composite walls involving several materials. For example, a domestic refrigerator has a sheet metal outside casing, a moulded plastic inner shell, with the space between filled with insulation. Taking a cross-section through any of the walls would reveal a composite structure of sheet metal, insulation and sheet plastic bonded together.

To analyse heat transfer situations involving fluids and composite walls requires the evaluation of an overall heat transfer coefficient, but one requiring a rather more complex relationship than that given in equation (12.5). The most straightforward way of evaluating such an overall heat transfer coefficient is by using an electrical resistance analogy for the thermal resistance of each separate mode of heat transfer.

With convection, the rate of heat transfer is a function of h_c, the convective heat transfer coefficient. For a given temperature difference between the surface and the fluid, the higher the value of h_c the higher the rate of heat transfer. Alternatively, it could be reasoned the higher the value of h_c the lower the 'resistance' to heat transfer. It is, therefore, possible to define the thermal resistance of a convective process as

$$R \text{ (convection)} = \frac{1}{h_c}$$

The thermal resistance for conduction can be thought of as **reducing** the higher the thermal conductivity becomes, but **increasing** directly with the thickness of the material:

$$R \text{ (conduction)} = \frac{x}{k}$$

Using these definitions of thermal resistance, equation (12.5) for the overall heat transfer coefficient can be expressed in the form

$$\frac{1}{U} = \frac{1}{h_{c\,w}} + \frac{x}{k} + \frac{1}{h_{c\,a}}$$

$$R_o = R_w + R_{wall} + R_a$$

where R_o is the overall thermal resistance equivalent to $1/U$. The overall thermal resistance can be found by summing the individual thermal resistances in the same way as electrical resistances operating in series. The combined convection and conduction situation shown in Figure 12.4 can now be defined in terms of the individual thermal resistances, as given in Figure 12.5.

Where a more complex situation than that given in Figure 12.5 is encountered, the same approach can be used by summing the thermal resistances in series:

$$R_o = R_1 + R_2 + R_3 \ldots \tag{12.7}$$

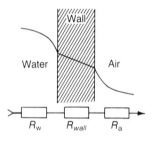

Figure 12.5 Resistance analogy for heat transfer.

Example 12.3

A domestic oven has a heat transfer surface area of 1.5 m². If the temperature inside the oven is maintained at 200°C when the ambient temperature is 20°C, calculate the heat lost from the oven.

Assume the walls of the oven to consist of an outside steel skin of 1.6 mm thickness, insulation of 25 mm thickness and an inner steel skin of 2 mm thickness. Take the values of thermal conductivity as k (steel) = 40 W/mK, k (insulation) = 0.05 W/mK. Take the convective heat transfer coefficient on the inside and outside surfaces to be 10 W/m²K.

Conceptual model

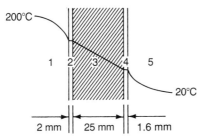

Analysis — the overall heat transfer coefficient can be evaluated using the resistance analogy defined by equation (12.7):

$$R_o = R_1 + R_2 + R_3 + R_4 + R_5$$

where

$$R_1 = 1/h_{c_1} = 1/10 = 0.1$$

$$R_2 = x_2/k_2 = \frac{2}{1000 \times 40} = 0.00005$$

$$R_3 = x_3/k_3 = \frac{25}{1000 \times 0.05} = 0.5$$

$$R_4 = x_4/k_4 = \frac{1.6}{1000 \times 40} = 0.00004$$

$$R_5 = 1/h_{c_5} = 1/10 = 0.1$$

Therefore,

$$R_o = 0.70009 \text{ m}^2\text{K/W}$$

and

$$U = \frac{1}{R_o} = \frac{1}{0.7} = 1.43 \text{ W/m}^2\text{K}$$

Substituting in equation (12.6)

$$Q = UA \, \Delta T = 1.43 \times 1.5 \times (200 - 20)$$
$$= 386.1 \text{ W}$$

12.4   MEAN TEMPERATURE DIFFERENCE

The situations discussed so far have had constant temperatures that result in a constant overall temperature difference. However, in heat exchangers, such as a car radiator, one fluid is being cooled by another fluid. In the case of a car radiator the jacket water is being cooled by air. As a result, there is a reduction in the temperature of the water and an increase in the temperature of the air. The change in temperatures across a heat exchanger have already been defined in section 5.7, using the steady flow energy equation.

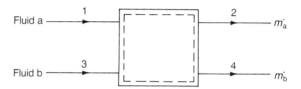

Figure 12.6 A heat exchanger.

Figure 12.6 shows a schematic diagram of a heat exchanger. There are two fluids 'a' and 'b' with two different flow rates, $\dot{m}_a$ and $\dot{m}_b$. Assuming that there is no change of phase within the fluids, they remain as liquids or gases, and that all the heat transfer takes place between the two fluid streams, the change of temperature can be found from equation (5.14):

$$(\dot{m}\, C_P)_a\,(T_1 - T_2) = (\dot{m}\, C_P)_b\,(T_4 - T_3)$$

Under these conditions it is still possible to evaluate the heat transfer performance using equation (12.6) with the temperature difference between the fluids now being a 'mean temperature difference':

$$Q = U\, A\, \Delta T_m \tag{12.8}$$

where ΔT_m is the mean temperature difference and the rate of heat transfer Q can be evaluated from either

$$(\dot{m}\, C_P)_a\,(T_1 - T_2)$$

or

$$(\dot{m}\, C_P)_b\,(T_4 - T_3).$$

Equation (12.8) represents a general relationship for all types of heat exchangers, but the calculation of the appropriate mean temperature difference depends on the actual values of the temperatures and the geometry of the particular heat exchanger.

12.4.1   Arithmetic mean temperature difference

Where the temperature changes in the two fluid flows of a heat exchanger are small, with relation to the difference between the entry temperatures, the mean temperature difference can be found by using an arithmetic mean value. Taking the temperatures as defined in Figure 12.6 with 'a' being the 'hot' fluid and 'b' being the cooling fluid, the arithmetic mean temperature difference is

$$\Delta T_m = \frac{(T_1 + T_2) - (T_3 + T_4)}{2} \tag{12.9}$$

This is valid for situations such as central heating and car radiators, where the water enters at a much higher temperature than the surrounding air and the difference between the entry temperatures dominates the situation. Figure 12.7 illustrates both types of radiator as open systems with the entry and leaving temperatures defined.

Figure 12.7(a) shows a diagram of a central heating radiator. The radiator consists of a tall narrow metal container with water flowing inside and air flowing outside. The hot water enters the radiator at the top and then sinks to the bottom as it cools. The water can be assumed to enter at T_1 and leave at T_2. The movement of the air will be adjacent to the radiator and the boundary defines a realistic limit to the heat exchanger. Air will rise against the radiator surface as it gets warmer than the surrounding air. The air can be assumed to enter at T_3 and leave at a higher temperature, T_4.

Both the water and air move due to the buoyancy associated with temperature changes in the fluid. Since the flow is by natural means the convection from the water to the inner surface of the radiator and the convection from the outer surface to the air are termed 'natural convection'.

By comparison, the flow of water and air through a car radiator, as shown in Figure 12.7(b), are 'forced' flows. The water is circulated through the radiator by a pump. The air is forced through the radiator by the forward movement of the car, possibly assisted by a fan. Under these circumstances, the convective processes are termed 'forced convection'. The type of convection determines the velocity of the flows and, therefore, the overall heat transfer coefficient, but

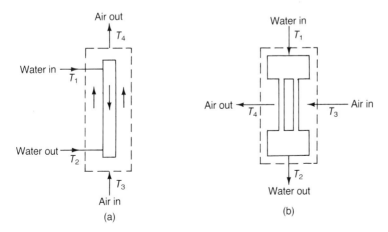

Figure 12.7 Central heating and car radiators.

has no effect on the mean temperature difference. It is the quantity of the flows that determines the temperature changes within the fluids and governs the mean temperature difference.

Before leaving the discussion of heat exchangers where the arithmetic mean temperature applies, it should be noted that the central heating radiator and car radiator represent two different configurations. The central heating radiator shown in Figure 12.7(a) has both the water and the air flowing in line but in opposite directions. This is termed a 'counter-flow' arrangement. The car radiator shown in Figure 12.7(b) has the direction of the air at right angles to the water flow. This is termed a 'cross-flow' arrangement.

Example 12.4

A central heating radiator has 3 l/min of of water entering at 80°C and leaving at 70°C. The air adjacent to the radiator is heated from 20°C to 25°C. Calculate the required surface area assuming that the overall heat transfer coefficient is the same as for example 12.2. Assume C_P for water to be 4.2 kJ/kg K.

Conceptual model

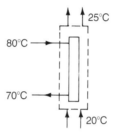

Analysis — the area can be evaluated using the general heat exchanger equation, (12.8):

$$Q = U \, A \, \Delta T_m$$

From example 12.2

$$U = 9.9 \text{ W/m}^2\text{K}$$

Assuming that the arithmetic mean temperature difference is valid,

$$\Delta T_m = \frac{(80 + 70) - (20 + 25)}{2} = 52.5 \text{ K}$$

The rate of heat transfer can be found from the flow rate on the water side:

$$Q = (m' \, C_P)_w \, (T_1 - T_2)$$

where

$$m' = 3/60 = 0.05 \text{ kg/s}$$
$$Q = 0.05 \times 4200 \times (80 - 70) = 2100 \text{ W}$$

Therefore,

$$2100 = 9.9 \times A \times 52.5$$
$$A = 4.04 \text{ m}^2$$

12.4.2 Log mean temperature difference

Where the variations of temperature within the fluids are large compared to the difference between the entry temperatures, it may not be appropriate to use an arithmetic mean temperature difference. In such situations, a more relevant form of mean temperature difference must be derived.

Consider a simple counter-flow heat exchanger consisting of two concentric tubes, as shown in Figure 12.8. The hot fluid 'a' flows through the inner tube and the cooling fluid 'b' flows in the opposite direction through the annular space formed between the inner and outer tubes. Figure 12.8 also shows the variation of temperatures within the heat exchanger. The dotted line representing the boundary of the open system has not been included in the diagram for the sake of clarity. It will be clear that the outer tube represents the boundary of the heat exchanger. It is assumed that there is **no heat transfer between the outer tube and the surroundings.** All heat transfer takes place across the inner tube between fluids 'a' and 'b'.

The heat exchanger will have a total surface area of A. If a very small element of the heat exchanger is considered, its area is dA and the heat transfer for this element will be

$$dQ = U \, \Delta T \, dA$$

where the temperature difference is

$$\Delta T = T_a - T_b$$

Differentiating the temperature difference gives

$$d(\Delta T) = d(T_a - T_b)$$
$$= dT_a - dT_b$$

$$= \frac{dQ}{(\dot{m} C_P)_a} - \frac{dQ}{(\dot{m} C_P)_b}$$

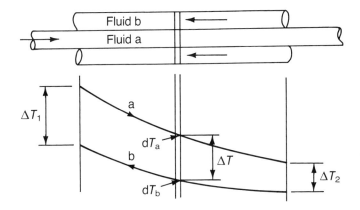

Figure 12.8 Counter-flow heat exchanger.

Re-arranging, and integrating for the whole heat exchanger,

$$\frac{\Delta T_2 - \Delta T_1}{Q} = \frac{1}{(\dot{m}\,C_P)_a} - \frac{1}{(\dot{m}\,C_P)_b} \tag{12.10}$$

Alternatively,

$$\frac{\mathrm{d}(\Delta T)}{\mathrm{d}Q} = \frac{\mathrm{d}(\Delta T)}{U\,\Delta T\,\mathrm{d}A} = \frac{1}{(\dot{m}\,C_P)_a} - \frac{1}{(\dot{m}\,C_P)_b}$$

which gives

$$\frac{\mathrm{d}(\Delta T)}{\Delta T} = \left(\frac{1}{(\dot{m}\,C_P)_a} - \frac{1}{(\dot{m}\,C_P)_b} \right) U\,\mathrm{d}A$$

Integrating for the whole heat exchanger,

$$\ln\frac{\Delta T_2}{\Delta T_1} = \left(\frac{1}{(\dot{m}\,C_P)_a} - \frac{1}{(\dot{m}\,C_P)_b} \right) UA \tag{12.11}$$

Equations (12.10) and (12.11) can be combined to give

$$Q = U\,A\,\Delta T_m$$

where the mean temperature difference is expressed as

$$\Delta T_m = \frac{\Delta T_2 - \Delta T_1}{\ln\frac{\Delta T_2}{\Delta T_1}} \tag{12.12}$$

This is termed the 'log mean temperature difference' and is true for any magnitude of terminal temperature differences ΔT_1 and ΔT_2, except for the case where $\Delta T_1 = \Delta T_2$. Under this situation, both $\Delta T_2 - \Delta T_1$ and $\ln(\Delta T_2/T_1)$ are zero and the log mean temperature difference gives an indeterminate result. For this limiting case it can be shown that $\Delta T_m = \Delta T_1 = \Delta T_2$ using the arithmetic mean temperature difference.

Example 12.5

Calculate the surface area for the radiator defined in example 12.4, using a log mean temperature difference instead of an arithmetic mean temperature difference.

Conceptual model — example 12.4.

Analysis — from example (12.4)

$$U = 9.9 \text{ W/m}^2\text{K}$$
$$Q = 2100 \text{ W}$$

The log mean temperature difference can be found from equation (12.12):

$$\Delta T_m = \frac{\Delta T_2 - \Delta T_1}{\ln\frac{\Delta T_2}{\Delta T_1}}$$

Taking the water inlet for ΔT_1,

$$\Delta T_1 = 80 - 25 = 55 \text{ K}$$
$$\Delta T_2 = 70 - 20 = 50 \text{ K}$$

Therefore,

$$\Delta T_m = \frac{50 - 55}{\ln\frac{50}{55}} = 52.46 \text{ K}$$

The new surface area is

$$A = 2100/9.9 \times 52.46 = 4.04 \text{ m}^2$$

Note — irrespective of which terminal temperature difference is taken as ΔT_1, the answer for ΔT_m would be the same.

It will be noted that for this situation the values of the arithmetic mean and the log mean temperature differences are the same, indicating that the arithmetic mean temperature difference is sufficiently accurate for this analysis.

12.4.3 Comparison between the mean temperature differences

The obvious question that arises from the foregoing discussion is when to use a log mean value of temperature difference instead of the simpler form of arithmetic mean. The answer is when the value of the arithmetic mean temperature difference involves too great an error. This can be seen by comparing the different values of mean temperature difference for various terminal difference ratios.

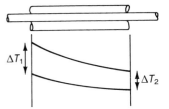

Figure 12.9 Counter-flow temperature distribution.

Consider a simple counter-flow heat exchanger as shown in Figure 12.9. The temperatures within the heat exchanger are defined in terms of the terminal temperature differences, ΔT_1 and ΔT_2.

The log mean temperature difference (LMTD) for such a situation can be found using equation (12.12). The arithmetic mean temperature difference (AMTD) can be expressed as:

$$\Delta T_m = \frac{\Delta T_1 + \Delta T_2}{2} \qquad (12.13)$$

A comparison of values of LMTD and AMTD is listed below for a range of ratios between ΔT_1 and ΔT_2.

ΔT_1	ΔT_2	LMTD	AMTD
100	90	94.9	95
100	75	86.9	87.5
100	50	72.1	75
100	25	54.1	62.5
100	10	39.0	55

From the above table it can be seen that where the terminal temperature differences vary by more than, say, 50% the use of the arithmetic mean is inadequate for calculating the temperature difference and the log mean temperature difference must then be used.

12.5 HEAT EXCHANGERS

Heat exchangers are devices in which energy is transferred from one fluid to another. Central heating and car radiators are two types of heat exchanger that have been discussed in connection with the mean temperature difference. They serve to show that heat exchangers come in different forms. Heat exchangers are categorized depending on the relative directions of the two fluid flows. Where the two fluids flow along the same axis, the heat exchanger is termed 'in-line'. Where the two fluids flow at right angles to each other, the heat exchanger is termed 'cross-flow'.

12.5.1 In-line heat exchangers

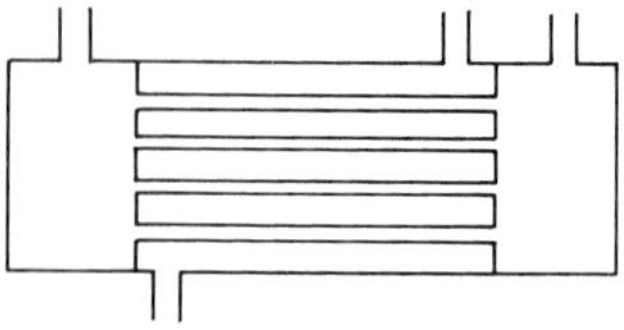

An in-line heat exchanger may be of a simple form of construction, consisting of two concentric tubes as shown in Figure 12.8. Alternatively, there may be several tubes within a much larger tube or shell. A shell and tube heat exchanger is shown in Figure 12.10.

Figure 12.10 Shell and tube heat exchanger.

Shell and tube heat exchangers are widely employed for the transfer of energy between two fluids or where there is a change of phase on the shell side. Steam boilers of the type where the hot gases from combustion flow through the tubes are typical of the configuration shown in Figure 12.10. Similarly, condensers used with steam power plants have steam condensing on the outside of the tubes as a result of cooling water flowing through the tubes.

Shell and tube heat exchangers without a change of phase can have one fluid flowing in one direction and the other flowing in the opposite direction,

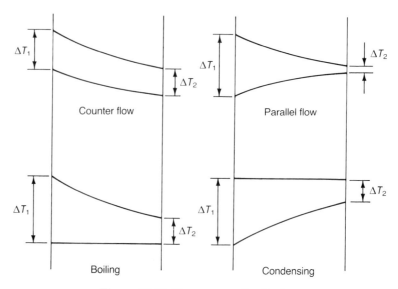

Figure 12.11 Temperature distributions.

a counter-flow arrangement. Occasionally, shell and tube heat exchangers have both fluids entering at the same end and flow in parallel through the exchanger. Such a heat exchanger is termed a 'parallel-flow' arrangement. The temperature distributions for the various arrangements are shown in Figure 12.11.

For any of the temperature distributions shown in Figure 12.11, the terminal temperature differences ΔT_1 and ΔT_2 can be defined and the mean temperature difference found using either equation (12.12) or (12.13).

Example 12.6

Oil is to be heated from 20°C to 50°C using water entering at 80°C and leaving at 60°C. It is proposed to use a shell and tube heat exchanger. Which would be the better arrangement, counter-flow or parallel-flow? Assume the same overall heat transfer coefficient for each arrangement.

Conceptual model — the two arrangements can be modelled using the following temperature distributions

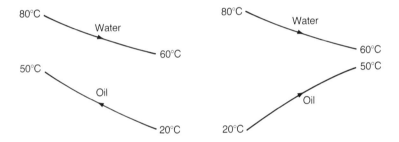

Analysis — the general relationship for both types of heat exchanger is given by equation (12.8):

$$Q = U \, A \, \Delta T_{\mathrm{m}}$$

Since Q and U are the same for both arrangements, it follows that the required surface area is a function of the mean temperature difference:

$$A \approx \frac{1}{\Delta T_{\mathrm{m}}}$$

The arrangement giving the highest value of ΔT_{m} results in the smallest amount of surface area, thereby reducing the cost of the heat exchanger.
For counter-flow

$$\Delta T_1 = 80 - 50 = 30 \text{ K}$$
$$\Delta T_2 = 60 - 20 = 40 \text{ K}$$

Since the terminal temperature differences are so close, ΔT_{m} can be found using the arithmetic mean from equation (12.13):

$$\Delta T_{\mathrm{m}} = \frac{\Delta T_1 + \Delta T_2}{2} = \frac{30 + 40}{2} = 35 \text{ K}$$

For parallel-flow

$$\Delta T_1 = 80 - 20 = 60 \text{ K}$$
$$\Delta T_2 = 60 - 50 = 10 \text{ K}$$

Using equation (12.12) to find the log mean temperature difference

$$\Delta T_m = \frac{10 - 60}{\ln \frac{10}{60}} = 27.9 \text{ K}$$

It follows that the counter-flow arrangement has the highest mean temperature difference and would require only 80% of the surface area of the parallel-flow arrangement.

12.5.2   Comparison between counter-flow and parallel-flow

The result from example 12.6 is typical of any heat exchanger, a counter-flow arrangement will always give a reduction in the surface area compared to a parallel-flow arrangement. In addition, a counter-flow arrangement can be used to cool a fluid to a temperature below the outlet temperature of the cooling fluid, which is impossible with a parallel-flow arrangement. With these advantages for the counter-flow arrangement, the obvious question that arises is why the parallel-flow arrangement is ever used, albeit only occasionally.

To answer this question it is necessary to not only consider the temperature variation in the two fluids but also to consider the variation of the temperature in the wall between the two fluids. Referring to the situation shown in Figure 12.4, the wall temperature will be intermediate between the temperatures of the two fluids at any particular position.

Figure 12.12 shows the variation of the wall temperature for a counter-flow and parallel-flow heat exchanger. It will be seen that the wall temperature varies more markedly for a counter-flow arrangement. Where a heat exchanger is being used with very high temperature fluids, there could be an advantage in using a parallel-flow arrangement if the lower wall temperature results in the use of less expensive materials. Alternatively, some chemical processes use parallel-flow heat exchangers to limit the wall temperature in which a particular chemical might be in contact.

12.5.3   Cross-flow heat exchangers

Cross flow can be introduced on the shell side of a shell and tube heat exchanger by the use of baffles. This is done to enhance the flow and, therefore,

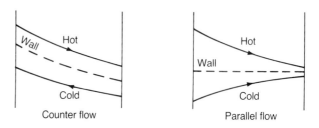

Figure 12.12 Variation of wall temperature.

the surface heat transfer coefficient on the shell side. However, this modification **does not** change the temperature distribution. A cross-flow heat exchanger is one in which the whole of the flow of one fluid is at right angles to the other fluid flow. As such, the majority of cross-flow heat exchangers use air as the outside fluid, either as a cooling medium or for space heating.

In the case of air being used as a cooling medium a car radiator is just one example. Many air cooled heat exchangers are used in industry for cooling process fluids. Alternatively, space heaters that can be free mounted, or as part of an air conditioning system, employ either steam or hot water inside the tubes to raise the temperature of the air flowing over the outside of the tubes. However, the main disadvantage of using air as a heat transfer fluid is the low convective heat transfer coefficient, particularly with respect to either water or condensing steam.

Assuming that the area is approximately the same for the outside of the tubes as for the inside, and ignoring the thermal resistance across the tube wall, the overall heat transfer coefficient is governed by the air-side performance. Taking typical convective heat transfer coefficient values of 50 W/m^{2}K on the air side and 2500 W/m^{2}K on the water side, the overall heat transfer coefficient is $1/(1/50 + 1/2500) = 49.02$ W/m^{2}K, which differs by less than 2% from that on the air side.

In such a situation the heat transfer analysis can be simplified by taking $U \approx h_{c_{air}}$. However, this results in an inefficient design as the excellent heat transfer performance of the water side is not fully utilised. It is possible to compensate for the poor heat transfer on the air side by increasing the effective surface area on that side with relation to the area on the water side. This can be done by adding fins to the air side. A typical example of an individual finned tube is shown in Figure 12.13.

In the case of finned tubes, the difference between the surface areas for the two fluids means that the general relationship for a heat exchanger equation (12.8) can be used but, instead of evaluating the overall heat transfer

Figure 12.13 Finned tube.

coefficient on its own, the product of UA is used. Taking the area on the air side as A_a and that on the water side A_w and ignoring the thermal resistance across the tube wall, the effective product of the overall heat transfer coefficient and surface area is

$$\frac{1}{UA} = \frac{1}{A_a h_{c_a}} + \frac{1}{A_w h_{c_w}} \qquad (12.14)$$

Clearly, equation (12.14) is valid for any finned tube heat exchanger working with fluids other than water and air.

Example 12.7

Two finned tube radiators are mounted on either side of a diesel-electric locomotive in order to dissipate a total of 2000 kW from the jacket cooling water. The pressurized water enters the radiators at 100°C and is cooled to 80°C with air entering at 30°C and leaving at 50°C. The radiators employ finned tube with an area of 0.05 m² on the water side and 0.5 m² on the air side for each metre length. Find the length of tube required in each radiator if the convective heat transfer coefficients are 50 W/m²K for air and 2000 W/m²K for water. Ignore the thermal resistance across the tube wall.

Conceptual model — for one radiator

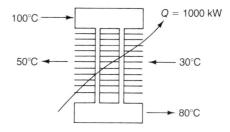

Analysis — taking one radiator and assuming that the mean temperature difference can be evaluated using an arithmetic mean,

$$\Delta T_m = \frac{(100+80)-(50+30)}{2} = 50 \text{ K}$$

Applying this value in the general equation, (12.8),

$$Q = U \, A \, \Delta T_m$$
$$1000 \times 10^3 = U \, A \times 50$$
$$UA = 20 \times 10^3 \text{ W/K}$$

Substituting values in equation (12.14):

$$\frac{1}{UA} = \frac{1}{A_a h_{c_a}} + \frac{1}{A_w h_{c_w}}$$

$$\frac{1}{20 \times 10^3} = \frac{1}{0.5 \times L \times 50} + \frac{1}{0.05 \times L \times 2000}$$

where L = tube length per radiator. Therefore,

$$0.00005 = \frac{0.04}{L} + \frac{0.01}{L}$$

$$L = \frac{0.05}{0.00005} = 1000 \, \text{m}$$

SUMMARY

In this chapter the three modes of heat transfer have been introduced and the application of combined convection and conduction has been discussed.

The key terms that have been introduced are:

convection
conduction
radiation
surface heat transfer coefficient
thermal conductivity
emissivity
overall heat transfer coefficient
thermal resistance
arithmetic mean temperature difference
log mean temperature difference
in-line heat exchanger
counter-flow
parallel-flow
cross-flow heat exchanger
finned tube

Key equations that have been introduced.

For convection between a surface and an adjacent fluid:

$$Q = h_c A \, \Delta T \qquad\qquad (12.1)$$

For conduction across a wall:

$$Q = (k/x)A \, \Delta T \qquad\qquad (12.2)$$

For combined convection and conduction for water and air across a plane wall:

$$\frac{1}{U} = \frac{1}{h_{c\,w}} + \frac{x}{k} + \frac{1}{h_{c\,a}} \qquad\qquad (12.5)$$

Thermal resistances in series:

$$R_o = R_1 + R_2 + R_3 + \ldots \qquad\qquad (12.7)$$

General relationship for a heat exchanger:

$$Q = U \, A \, \Delta T_m \qquad\qquad (12.8)$$

Log mean temperature difference:

$$\Delta T_m = \frac{\Delta T_2 - \Delta T_1}{\ln\frac{\Delta T_2}{\Delta T_1}}$$

(12.12)

Arithmetic mean temperature difference:

$$\Delta T_m = \frac{\Delta T_1 + \Delta T_2}{2}$$

(12.13)

Finned tube heat exchanger with water and air:

$$\frac{1}{UA} = \frac{1}{A_a h_{c_a}} + \frac{1}{A_w h_{c_w}}$$

(12.14)

PROBLEMS

1. A furnace has a refractory wall of 0.125 m thickness and thermal conductivity of 0.8 W/mK. If the inside surface is at 700°C and the outside surface is at 300°C, calculate:

   (a) the rate of heat transfer for a unit area of the wall;
   (b) the outside convective heat transfer coefficient if the surrounding air is at 25°C.

2. The furnace wall, defined in problem 1, is insulated on the outside by a 25 mm thickness of fibre-glass protected by sheet steel of 12 mm thickness. If the respective thermal conductivities k are 0.05 W/mK for fibreglass and 40 W/mK for steel find:

   (a) the new rate of heat transfer for a unit area of wall;
   (b) the outside temperature of the steel.

3. A window 1.5 m high by 1 m wide is double glazed with glass of 4 mm thickness and an air gap between of 7 mm thickness. On a day when the room temperature is 25°C and the outside air temperature is 0°C, calculate the heat loss across the window. Assume thermal conductivities of 1 W/mK for glass and 0.026 W/mK for the air gap, with convective heat transfer coefficients of 10 W/m²K both inside and outside.

4. A domestic freezer has an internal capacity of 1 m × 1 0.5 m × 0.5 m. The temperature inside is maintained at –5°C in ambient temperature of up to 35°C. Calculate the maximum rate of heat transfer to the freezer if the wall and door consist of:

5 mm plastic	$k = 1$ W/mK
15 mm insulation	$k = 0.05$ W/mK
2 mm steel	$k = 40$ W/mK

and the convective heat transfer coefficient is 8 W/m²K on both sides of the freezer.

5. An office block is constructed with a cavity wall. The wall is constructed of two brick sections, 0.11 m thick, separated by a cavity 50 mm wide. It is proposed to fill the cavity with foam plastic insulation to reduce the heat loss.

   Estimate the percentage saving in energy that could be achieved through this cavity insulation. Take k for brick = 0.7 W/mK and k for insulation = 0.05 W/mK. The convective heat transfer coefficient can be taken as 10 W/m²K for all vertical surfaces.

6. A water heater consists of a single tube with an electrical heating element wrapped around it. It can be assumed that the heating element maintains a constant temperature difference of 30 K between the inside surface of the tube and the water throughout its length. The flow rate is 2 l/min with water entering at 15°C and leaving at 65°C. Calculate the required length of tube if the diameter is 10 mm.

   Take the convective heat transfer coefficient on the water side to be 1200 W/m²K and C_P for water as 4.2 kJ/kg K.

7. A steam boiler consists of a shell and tube heat exchanger with the combustion gases flowing through the tubes and the boiling water on the shell side. The combustion gases enter at 2100 K and leave at 750 K. During a test at a steam pressure of 100 kPa the boiler produced 110 kg/s of saturated steam with water entering in a saturated condition.

   The boiler is designed for a steam pressure of 800 kPa. Estimate the rate of heat transfer at this pressure assuming that the overall heat transfer coefficient remains the same.

8. As part of an air-conditioning system, a finned tube heat exchanger is required to raise 10 kg/s of air from 8°C to 30°C using hot water entering at 80°C and leaving at 70°C. The type of finned tube has an inside area of 0.075 m² and an outside area of 0.6 m², for each metre length.

   Find the length of tube required in the heat exchanger if the convective heat transfer coefficients are air-side 40 W/m²K and water-side 1600 W/m²K. Ignore the thermal resistance across the tube wall. Take C_P for air as 1.005 kJ/kg K.

9. A heat exchanger has a gas flowing through the tubes, heated by steam condensing at 110°C on the outside of the tubes. The existing design has gas entering at 20°C and leaving at 50°C. It is necessary to increase the gas outlet temperature to 62°C and it has been proposed to modify the heat exchanger by inserting metal strips along the inside of each tube. Each tube would have one straight strip inside, having the same width as the tube diameter and in contact with the inside tube surface.

   Assuming that the overall heat transfer coefficient is not changed, would this proposed modification be satisfactory?

Outline solutions

CHAPTER 1

1. $W = \dfrac{\mathrm{Nm}}{\mathrm{s}} = \dfrac{\mathrm{kgm}}{\mathrm{s}^2}\dfrac{\mathrm{m}}{\mathrm{s}} = \dfrac{\mathrm{kgm}^2}{\mathrm{s}^3}$

2. $2.5\,\mathrm{h} = 150\,\mathrm{min} = 9000\,\mathrm{s}$, i.e. $9\,\mathrm{ks}$.

3. $P = \dfrac{gz}{v} + 100 = \dfrac{(9.81 \times 0.5)}{7.35 \times 10^{-5}} + \left(100 \times 10^2\right)$

   $\qquad = 166\,735\,\mathrm{Pa}$, i.e. $166.7\,\mathrm{kPa}$

4. $P = \left(\dfrac{gz}{v}\right)_{\mathrm{w}} + \left(\dfrac{gz}{v}\right)_{\mathrm{m}} = \dfrac{(9.81 \times 0.4)}{0.001} + \left(\dfrac{9.81 \times 0.75}{7.35 \times 10^{-5}}\right)$

   $\qquad = 104\,026\,\mathrm{Pa}$, i.e. $104\,\mathrm{kPa}$

5. $T = 37 + 273 = 310\,\mathrm{K}$

6. $M(CO_2) = 44,\ M(N_2) = 28$

   $m = (0.2 \times 44) + (0.8 \times 28) = 31.2\,\mathrm{kg}$

CHAPTER 2

1. An open system, as there is flow into and out of the radiator.

2. The potato is a closed system. The saucepan is an open system due to steam leaving to enter the atmosphere.

3. An open system with heat and work crossing the boundary.

4. Draw the boundary to encompass a fixed quantity of steam in the inlet pipe, cylinder and piston assembly.

5. Yes, draw the boundary around the boiler, engine and region of the atmosphere into which the steam exhausts.

6. $PE = mgz = 100 \times 10^3 \times 9.81 \times 5000 = 4905\,\mathrm{MJ}$

   $KE = \frac{1}{2}mV^2 = \frac{1}{2} \times 100 \times 10^3 \times (200)^2 = 2000\,\mathrm{MJ}$

7. Volume $= \pi(1)^2 \times 4 = 12.57$ m^3

   Mass $= 5000 + (12.57/0.0012) = 15\,475$ kg

   $KE = \frac{1}{2}mV^2 = \frac{1}{2} \times 15\,475 \times (15)^2 = 1.74$ MJ

8. $V = m'v/A = 75 \times 0.001/\pi(0.035)^2 = 19.5$ m/s

   $mgz = \frac{1}{2}mV^2$, $z = V^2/2g = (19.5)^2/2 \times 9.81 = 19.7$ m

9. $F = m'(V_0 - V_1) = 100(180 - 100) = 8000$ N

   Power $= \frac{1}{2}m'(V_0^2 - V_1^2) = \frac{1}{2}\,100(180^2 - 100^2) = 1120$ kW

CHAPTER 3

1. Yes, q is positive, $w = 0$, Δu is positive.

2. There is a supply of energy to the system through the wire from the mains, so the internal energy will increase.

3. $q_{net} = w_{net}$

   $40 + q_{21} = 30 - 45$, $q_{21} = -55$ kJ

4.

Process	q	w	u_1	u_2	Δu
a	20	15	35	40	5
b	8	−8	12	32	20
c	−10	2	−6	6	−12
d	25	15	18	28	10

5. $w = P(v_z - v_1)$

   $= 250(0.2 - 0.8) = -150$ kJ/kg

6. Consider the cylinder, pipe and balloon as a closed system.

   Work $= P$(volume 2 − volume 1)

   $= 100(0.01 - 0) = 1$ kJ

7. Substituting gives $b = 800$, $v_2 = 0.6$ m^3/kg

   From area under P − v diagram

   $$w = \left(\frac{600 + 200}{2}\right)(0.6 - 0.2) = 160 \text{ kJ/kg}$$

8. $w_{net} = w_{12} + w_{23} + w_{31}$

   $w_{12} = 160$ kJ/kg from question 6

   $w_{net} = 160 + 200(0.2 - 0.6) + 0 = 80$ kJ/kg

   $q_{net} = w_{net} = 80$ kJ/kg

9. $q_{net} = q_{12} + q_{23} + q_{34} + q_{41}$

   $q_{12} = T_1 \Delta s = 500 \times 2 = 1000$ kJ/kg

$q_{23} = 0$ (adiabatic)

$q_{34} = T_3 \, \Delta s = 300 \times -2 = -600$ kJ/kg

$q_{41} = 0$ (adiabatic)

$q_{net} = 1000 + 0 - 600 + 0 = 400$ kJ/kg

CHAPTER 4

1.   (a) $u = 762 + 0.75(2584 - 762) = 2128.5$ kJ/kg

    (b) $2000 = 763 + x(2778 - 763)$, $x = 0.614$

    (c) $T = 250 + 50(7 - 6.926)/(7.124 - 6.926) = 268.7$ K

2. $v_1 = 0.0235$ m^2/kg

    $x_2 = 0.0235/0.0554 = 0.424$

3. $h_1 = 605 + 0.6(2739 - 605) = 1885.4$ kJ/kg

$$h_2 = 2965 + \frac{30}{50}(3067 - 2965) = 3026.2 \text{ kJ/kg}$$

    $q = h_2 - h_1 = 3026.2 - 1885.4 = 1140.8$ kJ/kg

4. $s_2 - s_1 = (0.85 - 0.5)(0.9073 - 0.2956) = 0.2141$ kJ/kg K

    $q = T(s_2 - s_1) = 293 \times 0.2141 = 62.7$ kJ/kg

    also, pressure is constant so $q = h_2 - h_1$ which gives the same answer.

5. $q = 0$, $w = u_1 - u_2$

    $u_1 = 2600$ kJ/kg, $u_2 = 505 + 0.7(2530 - 505) = 1922.5$ kJ/kg

    $w = 2600 - 1922.5 = 677.5$ kJ/kg

6. $R = 8.314/44 = 0.189$ kJ/kg K

$$v = \frac{RT}{P} = \frac{0.189 \times 10^3 \times 293}{200 \times 10^3} = 0.277 \text{ m}^3/\text{kg}$$

    $m = 0.5/0.277 = 1.806$ kg

7. $q = h_2 - h_1 = 85$ kJ/kg

    $C_P = 85/(150 - 50) = 0.85$ kJ/kg K

$$\frac{v_2}{v_1} = \frac{T_2}{T_1} = \frac{423}{323} = 1.31$$

8. $R = 8.314/32 = 0.260$ kJ/kg K

$$v_2 = \frac{RT_1}{P_1} = \frac{0.260 \times 10^3 \times 373}{800 \times 10^3} = 0.121 \text{ m}^3/\text{kg}$$

$$q = w = P_1 v_1 \, \ln\frac{v_2}{v_1} = 800 \times 10^3 \times 0.121 \, \ln\frac{800}{200} = 134.2 \text{ kJ/kg}$$

9. $\gamma = (0.718 + 0.287)/0.718 = 1.4$

$$T_2 = T_1 \left(\frac{P_2}{P_1}\right)^{(\gamma-1)/\gamma} = 293(5)^{0.4/1.4} = 464 \text{ K}$$

$q = 0$, $w = u_1 - u_2 = C_v(T_1 - T_2) = 0.718(293 - 464) = -122.8 \text{ kJ/kg}$.

CHAPTER 5

1.   Steady flow, $q = 0$, $V_1 = V_2$

2. $v_1 = \dfrac{RT_1}{P_1} = \dfrac{297 \times 323}{150 \times 10^3} = 0.64 \text{ m}^3\text{/kg}$

     $v_2 = \dfrac{297 \times 473}{200 \times 10^3} = 0.702 \text{ m}^3\text{/kg}$

     $A_1 = \pi(0.075)^2 = 0.0177 \text{ m}^2$

     $A_2 = \pi(0.15)^2 = 0.0707 \text{ m}^2$

     $m^{\cdot} = V_1 A_1/v_1 = 100 \times 0.0177/0.64 = 2.766 \text{ kg/s}$

     $V_2 = 2.766 \times 0.702/0.0707 = 27.5 \text{ m/s}$

3. $v_2 = \dfrac{RT_2}{P_2} = \dfrac{287 \times 323}{101 \times 10^3} = 0.918 \text{ m}^3\text{/kg}$

     $m' = \dfrac{A V_2}{v_2} = \dfrac{\pi(25)^2 8}{10^6 \times 0.975} = 0.0171 \text{ kg/s}$

     $Q = m^{\cdot} C_P(T_2 - T_2) = 0.0171 \times 1005(50 - 20) = 515.9 \text{ W}$

4. $h_1 = 251 + 0.9(2609 - 251) = 2373 \text{ kJ/kg}$

     $h_2 = 251 \text{ kJ/kg}$

     $q = h_2 - h_1 = 251 - 2373 = -2122 \text{ kJ/kg}$

5. $T_2 = 2000(100/800)^{0.33/1.33} = 1194 \text{ K}$

     $v_2 = \sqrt{2C_P(T_1 - T_2)} = \sqrt{2 \times 1860(2000 - 1194)} = 1731.6 \text{ m/s}$

6. $h_1 = 92.02 \text{ kJ/kg}$

     $h_2 = 92.02 = 44.52 + x_2(242.83 - 44.52)$

     $x_2 = 0.24$

7. $h_1 = 3052 \text{ kJ/kg}$, $s_1 = 7.124 \text{ kJ/kg K}$

     $s_2 = 7.124 = 0.649 + x_2(8.149 - 0.649)$

     $x_2 = 0.863$, $h_2 = 192 + 0.863(2584 - 192) = 2256 \text{ kJ/kg}$

     $w = h_1 - h_2 = 3052 - 2256 = 796 \text{ kJ/kg}$

8. $h_2 = 670 + 0.85(2757 - 670) = 2444 \text{ kJ/kg}$

   $Q = \dot{m}(h_2 - h_1) = \dfrac{500}{3600}(2444 - 84) = 327.8 \text{ kw}$

9. $h_1 = 251 + 0.8(2609 - 251) = 2137 \text{ kJ/kg}$

   $\dot{m}_s(h_1 - h_2) = \dot{m}_w C_P(T_4 - T_3)$

   $200(2137 - 251) = \dot{m}_w 4.18(40 - 15); \ \dot{m}_w = 3610 \text{ kg/s}$

CHAPTER 6

1. At 1 mPa, $T_N = 179.9°C$, 452.9 K

   At 40 kPa, $T_L = 75.9°C$, 348.9 K

   $\eta = 1 - (348.9/452.9) = 0.23$, engine is not possible.

2. $Q_{in} = 8/0.28 = 28.6 \text{ MW}$

   $\dot{m}_f = \dfrac{28.6 \times 10^3}{46 \times 10^3} = 0.62 \text{ kg/s}$

3. $\eta = \dfrac{40 \times 3600}{12 \times 40000} = 0.3$

   $\eta_{Carnot} = \dfrac{500}{873} = 0.573$

   $\eta / \eta_{Carnot} = 0.3/0.573 = 0.52$

4. $Q \text{ rejected} = \left(\dfrac{1 - 0.3}{0.3}\right)30 = 70 \text{ kw}$

   $35 = \dot{m} \times 4.2(15), \ \dot{m}_{water} = 0.56 \text{ kg/s}$

5. $COP = 273/30 = 9.1$

   $Q_L = 50 \times 9.1 = 455 \text{ W}$

6. $T_H = 40°C, \ T_L = 15°C$

   $\eta = 313 - 288/313 = 0.08$

   $Q \text{ rejected} = 500\left(\dfrac{1}{0.08} - 1\right) = 5750 \text{ kw}$

7. At 2 MPa, $T_H = 212.4°C$, 485.4 K

   At 40 kPa, $T_L = 75.9°C$, 348.9 K

   $\eta = 1 - (348.9/485.4) = 0.28$

   $Q_{in} = 600/0.28 = 2143 \text{ MW}$

   $\dot{m}_f = \dfrac{2143 \times 10^3}{32 \times 10^3} = 67 \text{ kg/s}, 241 \text{ tonnes/h}$

8. $h_1 = 605$ kJ/kg, $h_2 = 605 + 0.9(2739 - 605) = 2526$ kJ/kg

$s_2 = 1.776 + 0.9(6.897 - 1.776) = 6.385$ kJ/kg K

$s_3 = 6.385 = 1.303 + x_3(7.359 - 1.303)$, $x_3 = 0.84$

$h_3 = 417 + 0.84(2675 - 417) = 2312$ kJ/kg

$$\eta = \frac{0.7(2526 - 2312)}{2526 - 605} = 0.078$$

9. $T_2 = 283(25)^{0.4/1.4} = 710$ K

$w_{rev} = 20 \times 1.03(710 - 283) = 8796$ kW

$w_a = 8796/0.84 = 10\,472$ kW, 10.5 MW

CHAPTER 7

1. (a) Improved; any increase in the temperature of the heat input improves efficiency.

   (b) Improved; superheating increases the enthalpy drop from 2 to 3.

2. $\eta = 1 - (298/338) = 0.118$

   $Q = w/\eta = 100/0.059 = 1695$ kW

3. $h_2 = 2769$ kJ/kg, $s_2 = 6.663$ kJ/kg K

   $6.663 = 1.026 + x_3(7.667 - 1.026)$, $x_3 = 0.849$

   $h_3 = 318 + 0.849(2636 - 318) = 2286$ kJ/kg

   $h_1 \approx h_4 = 318$ kJ/kg

   $$\eta = \frac{2769 - 2286}{2769 - 318} = 0.197$$

4. $h_2 = 3264$ kJ/kg, $s_2 = 7.464$ kJ/kg K

   $7.464 = 0.649 + x_3(8.149 - 0.649)$, $x_3 = 0.909$

   $h_3 = 192 + 0.909(2584 - 192) = 2366$ kJ/kg

   $h_1 \approx h_4 = 192$ kJ/kg

   $$\eta = \frac{3264 - 2366}{3264 - 192} = 0.292$$

   $Q_{in} = 50/0.292 = 171$ MW, $Q_{out} = 171 - 50 = 121$ MW

5. $h_2 = 3025$ kJ/kg, $s_2 = 6.768$ kJ/kg K

   $6.768 = 0.832 + x_2(7.907 - 0.832)$, $x_2 = 0.839$

   $h_3 = 251 + 0.839(2609 - 251) = 2229$ kJ/kg

$h_1 \approx h_4 = 251 \text{ kJ/kg}$

$\eta = \dfrac{3025 - 2229}{3025 - 251} = 0.287$

$Q_{in} = 200/0.287 = 697 \text{ MW}$

$m_f = \dfrac{697 \times 3600}{30 \times 10^3} = 83.6 \text{ tonnes/h}$

6. $COP = 263/(303 - 263) = 6.575$

$W = 250/6.575 = 38 \text{ W}$

7. $h_2 = 265.53 \text{ kJ/kg}, \ s_2 = 0.9051 \text{ kJ/kg K}$

$s_1 = 0.9051 = 0.1773 + x_1(0.9168 - 0.1773)$

$x_1 = 0.984.$

$h_1 = 44.52 + 0.984(242.83 - 44.52) = 239.66 \text{ kJ/kg}$

$h_4 = h_3 = 99.21 \text{ kJ/kg}$

$COP = \dfrac{239.66 - 99.21}{239.66 - 265.53} = 5.43$

8. $T_2 = 25 + 10 = 35°C$

$T_1 = -10 - 10 = -20°C$

$h_2 = 201.45 \text{ kJ/kg}, \ s_2 = 0.6839 \text{ kJ/kg K}$

$0.6839 = 0.0731 + x_1(0.7087 - 0.0731), \ x_1 = 0.961$

$h_1 = 17.82 + 0.961(178.73 - 17.82) = 172.45 \text{ kJ/kg}$

$h_4 = h_3 = 69.55 \text{ kJ/kg}$

$COP = \dfrac{172.45 - 69.55}{201.45 - 172.45} = 3.55$

9. $T_1 = 0 - 10 = -10°C$

$T_2 = 20 + 10 = 30°C$

$h_2 = 262.85 \text{ kJ/kg}, \ s_2 = 0.9057 \text{ kJ/kg K}$

$0.9057 = 0.1530 + x_1(0.9202 - 0.1530), \ x_1 = 0.981$

$h_1 = 38.04 + 0.981(239.91 - 38.04) = 236.07 \text{ kJ/kg}$

$h_4 = h_3 = 92.02 \text{ kJ/kg}$

$COP = \dfrac{236.07 - 92.02}{262.85 - 236.07} = 5.38$

$\dot{m} = 200/(236.07 - 92.02)10^3 = 1.388 \times 10^{-3} \text{ kg/s}$

CHAPTER 8

1. $\eta = 1 - (300/800) = 0.625$

    $w = 64 \times 0.625 = 40$ kJ/kg

    cycles $= 40/(40 \times 0.5) = 2/s$

2.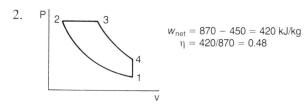

    $w_{net} = 870 - 450 = 420$ kJ/kg
    $\eta = 420/870 = 0.48$

3. $\eta = 1 - (1/6^{\,0.4}) = 0.51$

    $Q_{in} = 2 \times 10^{-3} \times 35\,000 = 70$ kW

    Power $= 70 \times 0.51 = 35.7$ kW

4. $T_2 = 290(8)^{0.4} = 666.2$ k

    $T_4 = 1500/(8)^{0.4} = 652.9$ K

    $w_{net} = 0.72(1500 - 666.2) - 0.72(652.9 - 290) = 339$ kJ/kg

    $Power = 0.002 \times 25 \times 339 = 16.9$ kW

5.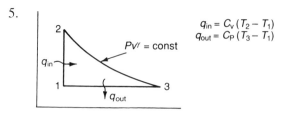

    $q_{in} = C_v\,(T_2 - T_1)$
    $q_{out} = C_P\,(T_3 - T_1)$

    $$\eta = \frac{C_v\,(T_2 - T_1) - C_P\,(T_3 - T_1)}{C_v\,(T_2 - T_1)} = 1 - \gamma\left(\frac{T_3 - T_1}{T_2 - T_1}\right)$$

    $$T_3 = \frac{v_3}{v_1}T_2 = 4 \times 290 = 1160 \text{ K}$$

    $T_2 = 1160\,(4)^{0.4} = 2020$ K

    $$\eta = 1 - 1.4\left(\frac{1160 - 290}{2020 - 290}\right) = 0.296$$

    for the Otto cycle $\eta = 1 - (1/4^{0.4}) = 0.426$.

6. $T_2 = 300(12)^{0.3/1.3} = 532.3 \text{ K}$

   $T_4 = 1100/(12)^{0.3/1.3} = 619.9 \text{ K}$

   $$\eta = \frac{1100 - 619.9 - 532.3 + 300}{1100 - 532.3} = 0.436$$

7. $T_2 = \sqrt{T_3 \times T_1} = \sqrt{1100 \times 300} = 574.5 \text{ K}$

   $T_4 = 574.5 \text{ K}$

   $W_{\text{net}} = 0.9(1100 - 574.5 - 574.5 + 300) = 225.9 \text{ kJ/kg}$

   $$\eta = \frac{1100 - 574.5 - 574.5 + 300}{1100 - 574.5} = 0.478$$

8. $T_2 = 293(14)^{0.4/1.4} = 622.8 \text{ K}$

   $T_4 = 1250/(14)^{0.4/1.4} = 588.1 \text{ K}$

   $W_{\text{net}} = 1.005(1250 - 588.1 - 622.8 + 293) = 333.8 \text{ kJ/kg}$

   $$\eta = \frac{1250 - 588.1 - 622.8 + 293}{1250 - 622.8} = 0.529$$

9. $T_2 = 290(16)^{0.38/1.38} = 622.3 \text{ K}$

   $T_4 = 1300/(16)^{0.38/1.38} = 605.9 \text{ K}$

   $$\eta = \frac{1300 - 605.9 - 622.3 + 290}{1300 - 622.3} = 0.534$$

   $Q_{\text{in}} = 5/0.534 = 9.36 \text{ MW}$

   $\dot{m}_{\text{f}} = 9.36 \times 10^3/44\,000 = 0.27 \text{ kg/s}$

   $W_{\text{comp}}/W_{\text{turb}} = (T_2 - T_1)/(T_3 - T_4)$
   $= (622.3 - 290)/(1300 - 605.9) = 0.48.$

CHAPTER 9

1. $T_2 = 293(12)^{0.4/1.4} = 595.9 \text{ K}$

   $T_4 = 1150/(12)^{0.4/1.4} = 565.4 \text{ K}$

   $$\eta = \frac{0.87(1150 - 565.4) - (595.9 - 293)}{1150 - 595.9} = 0.372$$

   $W = 10 \times 1.005(0.87(1150 - 565.4) - 595.9 + 293) = 2067 \text{ kW}$

2. $T_2 = 595.9 \text{ K}$

   $$0.82 = \frac{595.9 - 293}{T_{2a} - 293}, \qquad T_{2a} = 662.4 \text{ K}$$

$$T_4 = 565.4 \text{ K}$$

$$0.87 = \frac{1150 - T_{4a}}{1150 - 565.4}, \qquad T_{4a} = 641.4 \text{ K}$$

$$\eta = \frac{1150 - 641.4 - 662.4 + 293}{1150 - 662.4} = 0.285$$

$$W = 10 \times 1.005(1150 - 641.4 - 662.4 + 293) = 1399 \text{ kW}$$

3. $T_2 = 290(15)^{0.38/1.38} = 611.3 \text{ K}$

$$0.84 = \frac{611.3 - 290}{T_{2a} - 290}, \qquad T_{2a} = 672.5 \text{ K}$$

$$T_4 = 1250/(16)^{0.38/1.38} = 593.0 \text{ K}$$

$$0.89 = \frac{1250 - T_{4a}}{1250 - 593}, \qquad T_{4a} = 665.3 \text{ K}$$

$$\eta = \frac{1250 - 665.3 - 672.5 + 293}{1150 - 672.5} = 0.285$$

$$Q = 6/0.35 = 17.14 \text{ MW}$$

$$\dot{m}_f = 17.14/44 = 0.39 \text{ kg/s}$$

4. $T_2 = 288(18)^{0.37/1.37} = 628.6 \text{ K}$

$$628.6 - 288 = 1373 - T_4, T_4 = 1032.4 \text{ K}$$

$$T_5 = T_3 \, T_1/T_2 = 1373 \times 288/628.6 = 629.1 \text{ K}$$

$$\eta = \frac{1032.4 - 629.1}{1373 - 628.6} = 0.542$$

$$W = 25 \times 1.06(1032.4 - 629.1) = 10\,687 \text{ kW}$$

5. (a) $W = (5000 \times 150)/0.82 = 914.6 \text{ kW}$

$$Q = 914.6/0.4 = 2286.6 \text{ kW}$$

   (b) $\dot{m}_f = 2286.6/43000 = 0.0532 \text{ kg/s}$

6. $W = 20\,000 \times 150/0.5 = 1200 \text{ kW}$

$$T_2 = 288(19)^{0.4/1.4} = 612.1 \text{ K}$$

$$T_4 = 1100 - (612.1 - 288) = 775.9 \text{ K}$$

$$T_5 = 1100 \times 288/612.1 = 517.5 \text{ K}$$

$$\dot{m}_a = 1200/1.008(775.9 - 517.5) = 4.61 \text{ kg/s}$$

7. $T_2 = 293(14)^{0.38/1.38} = 606.0 \text{ K}$

$$606 - 293 = 1300 - T_4, T_4 = 987.0 \text{ K}$$

$T_5 = 1300 \times 293/606 = 628.5 \text{ K}$

$v_5 = \sqrt{2 \times 1050(987 - 628.5)} = 867.6 \text{ m/s}$

$F = 18 \times 867.6 = 15\ 617 \text{ N}$

8. $T_1 = 263 + \dfrac{(251)^2}{2 \times 1050} = 293 \text{ K}$

From problem 6: $T_2 = 606 \text{ K}$, $T_4 = 987 \text{ K}$

$T_5 = 1300 \times \dfrac{263}{606} = 564.2 \text{ K}$

$V_5 = \sqrt{2 \times 1050(987 - 564.2)} = 942.3 \text{ m/s}$

$F = 18(942.3 - 251) = 12\ 443 \text{ N}$

$\eta_{\text{prop}} = \dfrac{2 \times 251}{942.3 + 251} = 0.42$

9. $T_1 = 230 + \dfrac{(850)^2}{2 \times 1040} = 577.4 \text{ K}$

$T_3 = 1450 \times 230/577.4 = 577.6 \text{ K}$

$v_3 = \sqrt{2 \times 1040(1450 - 577.6)} = 1347.1 \text{ m/s}$

$F = 15\ (1347.1 - 850) = 7457 \text{ N}$

$\eta_{\text{prop}} = \dfrac{2 \times 850}{1347.1 + 850} = 0.77$

CHAPTER 10

1. $R = 0.4\left(\dfrac{8.314}{28}\right) + 0.25\left(\dfrac{8.314}{2}\right) + 0.35\left(\dfrac{8.314}{44}\right)$

     $= 1.224 \text{ kJ/kg K}$

(a) $N_2, P_1 = \left(\dfrac{m_1 R_1}{mR}\right)P = \dfrac{0.4 \times 0.297 \times 100}{1.224} = 9.7 \text{ kPa}$

    $H_2, P_2 = \dfrac{0.25 \times 4.157 \times 100}{1.224} = 84.9 \text{ kPa}$

    $CO_2, P_3 = \dfrac{0.35 \times 0.189 \times 100}{1.224} = 5.4 \text{ kPa}$

(b) $v = \dfrac{RT}{P} = \dfrac{1224 \times 290}{100 \times 10^3} = 3.55 \text{ m}^3/\text{kg}$

2. O_2, $P_1 = 400$ kPa, $R_1 = 8.314/32 = 0.260$ kJ/kg K

$$m_1 = \frac{P_1 \times \mathscr{V}}{RT} = \frac{400 \times 10^3 \times 1}{200 \times 323} = 4.763 \text{ kg}$$

N_2, $P_2 = 200$ kPa, $R_2 = 8.314/28 = 0.297$ kJ/kg K

$$m_2 = \frac{200 \times 10^3 \times 1}{297 \times 323} = 2.085 \text{ kg}$$

O_2, $4.763/(4.763 + 2.085) = 0.696$

N_2, $2.085/(4.763 + 2.085) = 0.304$

3. $C_P = 0.5 \times 0.85 + 0.3 \times 0.92 + 0.2 \times 14.2 = 3.541$ kJ/kg K

$q = C_P(T_2 - T_1) = 3.541(120 - 50) = 247.9$ kJ/kg

4. $R = 0.6\left(\dfrac{8.314}{28}\right) + 0.4\left(\dfrac{8.314}{44}\right) = 0.254$ kJ/kg K

$C_P = 0.6 \times 1.05 + 0.4 \times 0.94 = 1.006$ kJ/kg K

$\gamma = 1.006/(1.006 - 0.254) = 1.338$

$T_2 = 1200/(5)^{0.338/1.338} = 799.1$ K

$W = 1.006(1200 - 799.1) = 376.9$ kJ/kg

5. $P_{sat} = 3.166$ kPa

 (a) $0.7 = P_v/3.166$, $P_v = 2.216$ kPa
   $\omega = 0.622 \times 2.216/(101 - 2.216) = 0.014$ kg/(kg air)

 (b) $\omega = 0.0119 = 0.622\, P_v/(101 - P_v)$, $P_v = 1.896$ kPa
   $\phi = 1.896/3.166 = 0.6$.

6. $P_{v_1} = 0.8 \times 4.241 = 3.393$ kPa

$\omega_1 = 0.622 \times 3.393/(101 - 3.393) = 0.0216$ kg/(kg air)

$\omega_2 = 0.8 \times 0.0216 = 0.0173$ kg/(kg air)

$$0.0173 = 0.622 \frac{P_{sat_2}}{101 - P_{sat_2}}$$

$P_{sat_2} = 2.733$ kPa, i.e. $T_2 = 22.4°C$

7. $P_{v_1} = 0.6 \times 1.705 = 1.023$ kPa

$\omega_1 = 0.622 \times 1.023/(101 - 1.023) = 0.00636$ kg/(kg air)

$\omega_2 = 0.622 \times 4.241/(101 - 4.241) = 0.0273$ kg/(kg air)

Water lost/kg air $= 0.0273 - 0.00636 = 0.0209$ kg

8. $P_{v_1} = 0.7 \times 2.337 = 1.636$ kPa
   $\omega_1 = 0.622 \times 1.636/(100 - 1.636) = 0.0103$ kg/(kg air)

$\omega_2 = 0.622 \times 7.372/(100 - 7.372) = 0.0495$ kg/(kg air)

$\dot{m}_a = 1/(30 \times 60)(0.0495 - 0.0103) = 0.0142$ kg/s

9. $P_{v_1} = 0.5 \times 2.337 = 1.169$ kPa

$\omega_1 = 0.622 \times 1.169/(100 - 1.169) = 0.0074$ kg/(kg air)

$\omega_2 = 0.622 \times 5.62(100 - 5.62) = 0.0370$ kg/(kg air)

$h_1 = 20 + 0.0074(2501 + 1.82 \times 20) = 38.8$ kJ/kg air

$h_2 = 35 + 0.0370(2501 + 1.82 \times 35) = 129.9$ kJ/kg air

$Q = 12(129.9 - 38.8) = 1093.2$ kW

CHAPTER 11

1. $C_6H_6 + 7.5O_2 + 28.2 N_2 \rightarrow 6CO_2 + 3H_2O + 28.2N_2$

78 kg + 240 kg + 789.6 kg = 264 kg + 54 kg + 789.6 kg

A/F = (240 + 789.6)/78 = 13.2

2. Stoichiometric conditions,

$C_5H_{12} + 8O_2 + 30.1N_2 \rightarrow 5CO_2 + 6H_2O + 30.1N_2$

72 kg + 256 kg + 842.8 kg = 220 kg + 108 kg + 842.8 kg

Actual A/F = 1.2(256 + 842.8)/72 = 18.3

3. $C_{12}H_{24} + 36H_2O_2 \rightarrow 12CO_2 + 48H_2O$

168 kg + 1224 kg = 528 kg + 864 kg

H_2O_2/kg fuel = 1224/168 = 7.29 kg

4. $C_8H_{18} + 12.5O_2 + 47N_2 \rightarrow 8CO_2 + 9H_2O + 47N_2$

114 kg + 400 kg + 1316 kg = 352 kg + 72 kg + 1316 kg

$C_7H_{16} + 11O_2 + 41.36N_2 \rightarrow 7CO_2 + 8H_2O + 41.36N_2$

100 kg + 352 kg + 1158.1 kg = 308 kg + 144 kg + 1158.1 kg

stoichiometric A/F $= 0.8\left(\dfrac{1716}{114}\right) + 0.2\left(\dfrac{1510.1}{100}\right) = 15.06$

% excess air = ((18 − 15.06)/15.06)100 = 19.5%

5. $C_xH_y + z(O_2 + 3.76N_2) \rightarrow 7CO_2 + 8H_2O + 2O_2 + 48.9N_2$

$x = 7$, $y = 4$, $z = 13$

% excess air $= \left(\dfrac{13}{11} - 1\right) \times 100 = 18.2\%$

6. $C_2H_6 + 3.5O_2 \rightarrow 2CO_2 + 3H_2O$

    1 kmol + 3.5 kmol = 2 kmol + 3 kmol

    $\bar{h}_c = 2(-393\ 520) + 3(-241\ 820) - (-84\ 670)$
       $= -1\ 427\ 830\ kJ/kmol$

7. $C_8H_{18} + 12.5O_2 \rightarrow 8CO_2 + 9H_2O$

    1 kmol + 12.5 kmol = 8 kmol + 9 kmol

    $\bar{h}_c = 8(-393\ 520) + 9(-241\ 820) - (-249\ 950)$
       $= -5\ 074\ 590\ kJ/kmol$

    $LCV = 5\ 074\ 590/114 = 44\ 514\ kJ/kg$

8. $C_4H_{10} + x(O_2 + 3.76N_2) \rightarrow 4CO_2 + 5H_2O + (x - 6.5)O_2 + 3.76xN_2$
    1 kmol + x kmol + 3.76x kmol = 4 kmol + 5 kmol + x kmol − 6.5 kmol + 3.76x kmol
    $H_R = -126\ 140\ kJ$
    $H_p = 4(-314\ 084) + 5(-179\ 157) + x(51\ 679) - 6.5(51\ 679) + 3.76x(48992)$
    $0 = 2\ 488\ 034.5 + 235\ 888.9$

    $x = 10.55, \qquad A/F = \dfrac{1055\,(32) + 39.65\,(28)}{58} = 25$

9. $CH_4 + x(O_2 + 3.76N_2) \rightarrow CO_2 + 2H_2O + (x - 2)O_2 + 3.76xN_2$

    $\underbrace{1\ kmol + x\ kmol + 3.76x\ kmol}_{800\ K} = \underbrace{1\ kmol + 2\ kmol + x\ kmol - 2\ kmol + 3.76x\ kmol}_{1400\ K}$

    $H_R = -74\ 870 + x(15\ 838) + 3.76x\,(15045)$
       $= -74\ 870 + 72\ 407.2x$

    $H_P = (-337\ 617) + 2(-198\ 342) + x(36\ 956) - 2(36\ 956) + 3.76x\,(34941)$
       $= -808\ 213 + 168\ 334.2x$

    $0 = -808\ 213 + 168334.2x + 74870 - 72407.2x$

    $x = 7.64, \qquad \%\ \text{excess air} = \left(\dfrac{7.64}{2} - 1\right) = 282\%$

CHAPTER 12

1.   (a) $\dfrac{Q}{A} = \dfrac{0.8}{0.125}(700 - 300) = 2560\ W/m^2$

    (b) $h_c = 2560/(300 - 25) = 9.31\ W/m^2K$

2.   (a) $\dfrac{1}{U} = \dfrac{0.125}{0.8} + \dfrac{0.025}{0.05} + \dfrac{0.012}{40} + \dfrac{1}{9.31}, \qquad U = 1.31\ W/m^2$

$$\frac{Q}{A} = 1.31(700 - 25) = 884.3 \text{ W/m}^2$$

(b) $884.3 = 9.31(T - 25)$, $T = 120°C$

3. $\dfrac{1}{U} = \dfrac{1}{10} + \dfrac{0.004}{1} + \dfrac{0.007}{0.026} + \dfrac{0.004}{1} + \dfrac{1}{10}$, $\quad U = 2.1 \text{ W/m}^2 \text{ K}$

$Q = 2.1 \times 1.5(25 - 0) = 78.6 \text{ W}$

4. $\dfrac{1}{U} = \dfrac{1}{8} + \dfrac{0.005}{1} + \dfrac{0.015}{0.05} + \dfrac{0.002}{40} + \dfrac{1}{8}$, $\quad U = 1.8 \text{ W/m}^2 \text{ K}$

$Q = 1.8(4 \times 0.5 + 2 \times 0.25)(35 + 5) = 180 \text{ W}$

5. Cavity: $\dfrac{1}{U} = \dfrac{1}{10} + \dfrac{0.11}{0.7} + \dfrac{1}{10} + \dfrac{1}{10} + \dfrac{0.11}{0.7} + \dfrac{1}{10}$

$U = 1.4 \text{ W/m}^2 \text{ K}$

Insulated: $\dfrac{1}{U} = \dfrac{1}{10} + \dfrac{0.11}{0.7} + \dfrac{0.05}{0.05} + \dfrac{0.11}{0.7} + \dfrac{1}{10}$

U = 0.66 W/m² K, i.e. 53% saving

6. $Q = (2/60)4200. (65 - 15) = 5600 \text{ W}$

$A = 5600/(1200 \times 30) = 0.156 \text{ m}^2$

$L = 0.156/(\pi \times 0.01) = 4.97 \text{ m}^2$

7. At 100 kPa: $Q = 110 \times (2675 - 417) = 248.4 \text{ kW}$
$\Delta T_1 = 2100 - 99.6 \approx 2000 \text{ K}$, $\Delta T_2 = 750 - 99.6 \approx 650 \text{ K}$

$$\Delta T_m = \frac{650 - 2000}{\ln(650/2000)} = 1201 \text{ K}$$

At 800 kPa: $\Delta T_1 = 2100 - 170.4 \approx 1930 \text{ K}$, $\Delta T_2 = 750 - 170.4 \approx 580 \text{ K}$

$$\Delta T_m = \frac{580 - 1930}{\ln(580/1930)} = 1122.9 \text{ K}$$

$$Q = \frac{1122.9}{1201} \times 248.4 = 232.2 \text{ kW}$$

8. $Q = 10 \times 1.005 (30 - 8) = 221.1 \text{ kW}$

$\Delta T_m = ((80 + 70) - (8 + 30))/2 = 56 \text{ K}$

$UA = (221.1 \times 10^3)/56 = 3948.2 \text{ W/K}$

$$\frac{1}{3948.2} = \frac{1}{40L\,0.6} + \frac{1}{1600L\,0.075}, \qquad L = 197.4 \text{ m}$$

9. Required $\dfrac{Q_2}{Q_1} = \dfrac{(62-20)}{(50-20)} = 1.4$

$$\frac{A_2}{A_1} = \frac{(\pi D + 2D)}{\pi D} = 1.64$$

$\Delta T_{m_1} = ((110-20) + (110-50))/2 = 75 \text{ K}$
$\Delta T_{m_2} = ((110-20) + (110-62))/2 = 69 \text{ K}$

$$\frac{Q_2}{Q_1} = \frac{UA_2\Delta T_{m_2}}{UA_1\Delta T_{m_1}} = 1.64\,\frac{69}{75} = 1.51, \qquad \text{satisfactory}$$

Appendix A

A1 SATURATED WATER-STEAM PROPERTIES

Pressure (kPa)	Temp (°C)	Specific volume (m³/kg)	Internal energy (kJ/kg)		Enthalpy (kJ/kg)		Entropy (kJ/kg K)	
P	T_s	v_g	u_f	u_g	h_f	h_g	s_f	s_g
10	45.8	14.67	192	2437	192	2584	0.649	8.149
20	60.1	7.648	251	2456	251	2609	0.832	7.907
40	75.9	3.992	318	2476	318	2636	1.026	7.667
60	86.0	2.731	360	2489	360	2653	1.145	7.531
80	93.5	2.087	392	2498	392	2665	1.233	7.434
100	99.6	1.694	417	2508	417	2675	'1.303	7.359
200	120.2	.8856	505	2530	505	2707	1.530	7.127
400	143.6	.4623	605	2554	605	2739	1.776	6.897
600	158.8	.3156	669	2568	670	2757	1.931	6.761
800	170.4	.2403	720	2577	721	2769	2.046	6.663
1000	179.9	.1944	762	2584	763	2778	2.138	6.586
2000	212.4	.0996	907	2600	909	2799	2.447	6.340

The values of properties are based on those given in Rogers and Mayhew (1988).

The values of u_f and h_f are the same at low pressure due to the difference Pv_f being negligible.

The values are based on a datum of u_f and s_f being zero at a saturation temperature of 0.01°C. This temperature is called the triple point as all three phases of water; solid, liquid and vapour; can co-exist in equilibrium at this temperature at a pressure of 0.6112 kPa.

A2 SUPERHEATED STEAM PROPERTIES

Pressure (kPa) P	Temp (°C) T_s		Steam temperatures (°C)					
			150	200	250	300	350	400
100	99.6	h	2777	2876	2975	3075	3176	3278
		s	7.614	7.834	8.033	8.215	8.384	8.543
200	120.2	h	2770	2871	2971	3072	3174	3277
		s	7.280	7.507	7.708	7.892	8.063	8.221
400	143.6	h	2753	2862	2965	3067	3170	3274
		s	6.929	7.172	7.379	7.566	7.738	7.898
600	158.8	h		2851	2958	3062	3166	3270
		s		6.968	7.182	7.373	7.546	7.707
800	170.4	h		2840	2951	3057	3162	3267
		s		6.817	7.040	7.233	7.409	7.571
1000	179.9	h		2829	2944	3052	3158	3264
		s		6.695	6.926	7.124	7.301	7.464
2000	212.4	h			2904	3025	3137	3248
		s			6.547	6.768	6.957	7.126

The values of properties are based on those given in Rogers and Mayhew (1988).

Values of specific enthalpy h are given in kJ/kg and of specific entropy s are given in kJ/kg K.

Only values of h and s are presented since superheated steam is largely used in steam power cycles for which these two properties are sufficient for analysis.

A3   SATURATED REFRIGERANT PROPERTIES OF REFRIGERANT-12 (CF$_2$ Cl$_2$)

Temp (°C)	Pressure (kPa)	Specific volume (m^3/kg)	Enthalpy (kJ/kg)		Entropy (kJ/kg K)	
T_s	P	v_g	h_f	h_g	s_f	s_g
−20	150.9	.1088	17.82	178.73	.0731	.7087
−15	182.6	.0910	22.33	180.97	.0906	.7051
−10	219.1	.0766	26.87	183.19	.1080	.7020
−5	261.0	.0650	31.45	185.38	.1251	.6991
0	308.6	.0554	36.05	187.53	.1420	.6966
5	362.6	.0475	40.69	189.66	.1587	.6943
10	423.3	.0409	45.37	191.74	.1752	.6921
15	491.4	.0354	50.10	193.78	.1915	.6901
20	567.3	.0308	54.87	195.78	.2078	.6885
25	651.6	.0269	59.70	197.73	.2239	.6869
30	744.9	.0235	64.59	199.62	.2399	.6853
35	847.7	.0206	69.55	201.45	.2559	.6839
40	960.7	.0182	74.59	203.20	.2718	.6825

The values of properties are based on those given in Rogers and Mayhew (1988).
The values are based on a datum with h_f and s_f being zero at a saturation temperature of −40°C. The molecular weight M = 120.9.

A4   SATURATED REFRIGERANT PROPERTIES OF REFRIGERANT-134A (CH$_2$ FCF$_3$)

Temp (°C)	Pressure (kPa)	Specific volume (m^3/kg)	Enthalpy (kJ/kg)		Entropy (kJ/kg K)	
T_s	P	v_g	h_f	h_g	s_f	s_g
−20	132.7	.1447	25.24	234.08	.1036	.9286
−15	163.9	.1184	31.62	237.00	.1285	.9241
−10	200.5	.0978	38.04	239.91	.1530	.9202
−5	243.2	.0813	44.52	242.83	.1773	.9168
0	292.5	.0681	51.06	245.75	.2013	.9141
5	349.2	.0575	57.67	248.66	.2251	.9118
10	413.9	.0487	64.35	251.56	.2488	.9099
15	487.3	.0416	71.12	254.43	.2722	.9084
20	570.2	.0356	77.98	257.28	.2956	.9073
25	663.4	.0306	84.95	260.09	.3189	.9064
30	767.5	.0265	92.02	262.85	.3421	.9057
35	883.5	.0229	99.21	265.53	.3653	.9051
40	1012.2	.0199	106.53	268.12	.3886	.9046

The values of properties are based on those given in Thermodynamic Properties of HFA 134a (1990).
The values are based on a datum with h_f and s_f being zero at a saturation temperature of −40°C. The molecular weight M = 102.

Appendix B

B1 SATURATION PRESSURE OF WATER VAPOUR

Temp T (°C)	Saturated pressure P_{sat} (kPa)
5	0.872
10	1.227
15	1.705
20	2.337
25	3.166
30	4.241
35	5.620
40	7.372
45	9.577

The values of saturation pressure are based on the relationship given in Bull (1964):

$$\log P_{sat} = \frac{7.5T}{237.3 + T} + 2.7857$$

where the pressure is given in Pa.

B2 PSYCHROMETRIC CHART

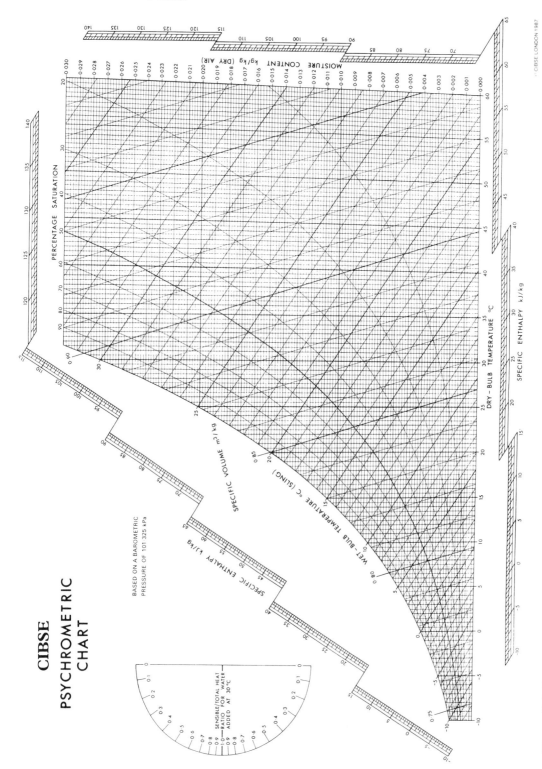

Figure B.2 Reproduced by permission of the Chartered Institution of Building Services Engineers. Pads of charts for calculation purposes may be obtained from CIBSE, 222 Balham High Road, London SW12 9BS.

Appendix C

C1 ENTHALPY OF FORMATION

Substance	Formula	State	$\overline{h}_f$ (kJ/kmol)
carbon dioxide	CO_2	gas	−393 520
water vapour	H_2O	gas	−241 820
methane	CH_4	gas	−74 870
ethane	C_2H_6	gas	−84 670
propane	C_3H_8	gas	−103 840
butane	C_4H_{10}	gas	−126 140
heptane	C_7H_{16}	liquid	−224 390
octane	C_8H_{18}	liquid	−249 950

The values of enthalpy of formation are quoted for standard reference conditions of 25°C and 1 atmosphere.

The values are based upon those given in Joint Army Navy Air Force (JANAF) Thermochemical Tables, 1971.

Values of enthalpy of formation for other substances can be found in Look and Sauer (1988).

C2 ENTHALPY OF GASES

T(K)	Oxygen (kJ/kmol)	Nitrogen (kJ/kmol)	Carbon dioxide (kJ/kmol)	Water vapour (kJ/mol)
298	0	0	−393 520	−241 820
400	3 028	2 972	−389 513	−238 365
600	9 249	8 895	−380 605	−231 316
800	15 838	15 045	−370 707	−223 820
1000	22 701	21 459	−360 118	−215 830
1200	29 758	28 110	−349 041	−207 323
1400	36 956	34 941	−337 617	−198 342
1600	44 269	41 913	−325 947	−188 933
1800	51 679	48 992	−314 084	−179 157
2000	59 189	56 156	−302 078	−169 065
2200	66 792	63 380	−289 951	−158 712
2400	74 484	70 661	−277 737	−148 139

The values of enthalpy are based upon those given in JANAF Thermochemical Tables, 1971.

The values of enthalpy are derived from the relationship

$$\overline{h} = \overline{h}_f + \Delta\overline{h}$$

where $\overline{h}_f$ is the enthalpy of formation at a temperature of 298 K and $\Delta\overline{h}$ is the increase in enthalpy associated with the change in temperature $(T - 298)$ K.

References and suggested reading

Allen, J.S. (1986) *The Newcomen Engine*, Black Country Museum, Dudley.

Bragg, S.L. (1962) *Rocket Engines*, George Newnes Ltd, London.

Bull, L.C. (1964) Design and Use of the New IHVE Psychrometric Chart, *Journal of the Institute of Heating and Ventilating Engineers*.

Carnot, S. (1960) *Reflection on the Motive Power of Heat and on Machines Fitted to Develop this Power*, Dover Publications, New York.

Cengal, Y.A. and Boles, M.A. (1989) *Thermodynamics – An Engineering Approach*, McGraw-Hill, New York.

Derry, T.K. and Williams, T.I. (1970) *A Short History of Technology*, Oxford University Press, Oxford.

Eastop, T.C. and McConkey, A. (1986) *Applied Thermodynamics for Engineering Technologists*, Longman, Harlow.

The International System (SI) Units, BS 3763, (1976), British Standards Institution.

The Jet Engine (1987) Rolls Royce plc, Derby.

Look, D.C. and Sauer, H.J. (1988) *Engineering Thermodynamics*, SI Edition, Chapman & Hall, London.

Massey B.S. (1989) *Mechanics of Fluids*, Chapman & Hall, London.

Newton, K. Steeds, W. and Garrett, T.K. (1989) *The Motor Vehicle*, Butterworths, London.

Rogers, C.F.C. and Mayhew, Y.R. (1988) *Thermodynamic and Transport Properties of Fluids*, Basil Blackwell, Oxford.

Simonson, J.R. (1988) *Engineering Heat Transfer*, Macmillan, London.

Thermodynamic Properties of HFA 134a (1990) ICI Chemicals and Polymers Ltd.

Van Wylen, G.J. and Sonntag, R.E. (1985) *Fundamentals of Classical Thermodynamics*, Wiley, New York.

Whitfield, P.R. (1975) *Creativity in Industry*, Penguin Books, Harmondsworth.

Whittle, F. (1953) *Jet*, Frederick Muller, London.

Index